The Math Dude's

Quick and Dirty Guide to Algebra

The Math Dude's

Quick and Dirty Guide to Algebra

JASON MARSHALL

ST. MARTIN'S GRIFFIN ❧ NEW YORK

www.stmartins.com

ISBN 978-0-312-56956-3

First Edition: July 2011

10   9   8   7   6   5   4   3   2   1

This book is dedicated to all the "math fans" of the world. And, in particular, to the great community of *The Math Dude* podcast listeners. Without your interest and support there would be math, but there would be no *Math Dude*.

Thank you!

CONTENTS

INTRODUCTION	1
What Can This Book Do for You?	1
How Should You Use This Book?	7
PROLOGUE: WHY MATH ISN'T AN AWFUL NERD	9
Basic Number Properties	10
Basic Arithmetic	11
Let the Game Begin!	13
Looking for Patterns in Numbers	17
Exponents and Perfect Squares	19
A Surprising Sequence of Numbers	20
Wrap-up	26
Final Exam	27

PART I: WHAT IS ALGEBRA, REALLY?

CHAPTER 1: TAKING ALGEBRA TO THE STREETS	33
The Secret Algebra You've Already Been Doing	34
Using Variables	38

Writing Equations 40

What's the Point of Algebra? 42

Square Roots 46

The Pythagorean Theorem 49

Why Algebra Matters in the Real World 56

Algebra in Your Backyard 62

Algebra Tutorial: How to Make a Graph 63
Challenge Problems 71
Final Exam 72

CHAPTER 2: ALGEBRA BASICS 76

What Are Variables? 77

How Do Variables Work? 82

What Are Algebraic Expressions? 86

Algebra Tutorial: The Order of Operations 88

Algebra Tutorial: Practice Problem 92

Intro to Equations 95

Halftime Recap 99

How You Should Think About Equations 106

Algebra Tutorial: How to Solve an Equation 107

Algebra Tutorial: How to Solve an Algebra Problem 123

Wrap-up 129

Final Exam 130

PART II: UNDERSTANDING ALGEBRA BETTER

CHAPTER 3: WALK THE NUMBER LINE 135

A Brief History of Numbers 136

Algebra and Decimal Numbers 141

Algebra and the Number Line 143

Absolute Values 144

Number Boot Camp   155

Linear Equations   164

Algebra Tutorial: How to Solve Single Variable
    Linear Equations   167

Linear Equations with Absolute Values   174

Algebra Tutorial: How to Solve Absolute
    Value Equations   176

Linear Inequalities   179

Algebra Tutorial: How to Solve Linear Inequalities   181

Wrap-up   193

Final Exam   194

**CHAPTER 4: ARITHMETIC 2.0: MATH WITH VARIABLES,
EXPONENTS, AND ROOTS**   197

Arithmetic 1.0: Preparing for Algebra   198

Doing Arithmetic with "Real" Numbers   201

Arithmetic 2.0 (Here Comes the Algebra)   214

Algebra Tutorial: How to Simplify Expressions   220

Math Properties   223

Algebra Tutorial: How to Combine Like Terms   238

Exponentiation   240

Roots   248

Exponentiation and Roots Combined   254

Irrational Exponents   256

Can You Remind Me of the Point of Algebra?   259

Final Exam   260

PART III: SOLVING ALGEBRA PROBLEMS

CHAPTER 5: POLYNOMIALS, FUNCTIONS, AND BEYOND   265

What Are Polynomials?   266

Evaluating Polynomials   274

Functions   279

Visualizing Polynomial Functions   287

How to Solve Problems with Polynomials   296

Algebra Tutorial: How to Find and Write
    the Equations of Lines   300

Equations of Horizontal and Vertical Lines   309

Systems of Equations   320

Algebra Tutorial: How to Solve a System of Equations   321

Systems of Inequalities   328

Algebra Tutorial: How to Solve a System of Inequalities   329

Challenge Problem #1   333

Final Exam   335

CHAPTER 6: THE ROOT OF THE PROBLEM   339

Challenge Problem #2   339

Roots of Polynomials   344

Factoring Polynomials   360

Algebra Tutorial: How to Factor Polynomials   372

Solving Quadratic Equations   383

Challenge Problem #3   401

Game Over?   409

Final Exam   410

The Math Dude's Solutions   413

Acknowledgments   475

Index   477

Introduction

Greetings, math fan!

I'm going to go out on a limb and guess that you're holding this book because you want or need to learn algebra. Or perhaps you just thought the cover looked nice and felt compelled to pick it up. Either way, I'm glad you did. And I think you'll be glad that you did too.

WHAT CAN THIS BOOK DO FOR YOU?

My offer is to teach you not just how to do algebra, but about the meaning of it too. Why should that be important to you? Well, let me give you an analogy. Though it's useful to learn to follow a couple of recipes that will help you prepare a few meals, eventually you'll grow tired of eating the same chicken stir-fry day after day after day. But if you instead learn about ingredients and all the ways you can combine and prepare them, then you'll have learned to cook. And with that knowledge will come a lifetime of lasagna and steak and risotto and lobster—in other words, you won't be doomed to a life of just chicken stir-fry.

That's exactly analogous to what we're going to do with algebra! When all is said and done, you're not just going to know how to do a couple of specific problems, or cook some classic "recipes"; you're going to understand how and why algebra works so that you have the ability to solve all the problems this book doesn't teach you to solve too—problems that you might see on the SATs, on a final exam, or just on your homework. That's the true test of whether or not you have really learned something. And that's exactly the test that this book is aimed at helping you pass.

WHO CARES ABOUT ALGEBRA?

I certainly do. And I hope that you do too. Maybe you're a middle school or high school student taking an algebra class. Or perhaps you're a (slightly embarrassed) parent of one of those kids that used to help with homework but now hides at the sound of their son's or daughter's approaching footsteps. Or maybe it's been a few years since your last math class and you need to refresh those memories for your new job. And, of course, there are those people who want to learn simply because they know that there's nothing better in life than discovering something new—or rediscovering something old. Maybe you're one of those lucky people.

In truth, it doesn't really matter which of those you are—or whether or not your description matches any of them. Because this book is here for each and every one of you. And so am I. Honestly, if you have questions while reading the book, need help understanding something, or just want to chat about math, you're more than welcome to write me . . . I'd love to hear from you. My e-mail address is mathdude@quickanddirtytips.com.

WHO IS THE MATH DUDE?

When I was young—before I learned algebra or much math at all—I remember thumbing through a dusty and ancient-looking algebra book I found in my great-grandmother's basement. I can remember looking at all the symbols and equations on the pages and thinking that it looked really awesome. I had no idea how it worked or what any of it meant, but I certainly wanted to know. Kind of strange, I know. But that's the kind of kid I was . . . and it's still the kind of person I am today—curious. I'm never satisfied to know only *that* something works because what I really want to know is *why* it works. To me, that's the good part!

Once I got a little older, I realized that the universe is a really big place . . . and that I wanted to know how it works too. So I studied astrophysics at UCLA, and eventually earned a Ph.D. from Cornell University. Which means that my background with math is not, as you might expect, as a mathematician or a math teacher, but is instead as a professional user of math. After a computer, a familiar programming language, and a cup of coffee or two, math is the most essential tool that I have to help me study the universe.

After graduate school, I made a few important discoveries. First, though I had always been good at math and wanted to know as much about it as possible, I must admit that I had never found it beautiful. But at some point curiosity (and the Internet) got the better of me, and I started reading about and playing with math . . . just like I had wanted to do as a kid. And what I discovered was truly exquisite. Oh, at some point in there I also made the discovery that I'm not half bad at explaining complicated things to people.

Which leads me directly to why you're reading these very words at this very moment. Because in January of 2010, I took action on the twin discoveries I'd made and became Quick and Dirty Tips' resident Math Dude. Each and every week since then I've been writing and hosting a new podcast episode of *The Math Dude's Quick and Dirty Tips to Make Math Easier.* iTunes even named it one of the best new podcasts of 2010! So, if you end up liking what you read here, I encourage you to join the legions of "math fans" and give the podcast a listen. You can find the show on iTunes or at http://mathdude.quickanddirtytips.com.

WHAT'S IN THIS GUIDE TO ALGEBRA?

If you're wondering if this book will cover *all* of algebra, the answer is no. No book can. But it does cover the big important topics that you'll see in a typical first course in algebra—usually called Algebra I. Which means that we will cover everything from linear equations to polynomials to how to solve quadratic equations.

HOW IS THIS GUIDE ORGANIZED?

In order to learn any subject, you have to practice. It's not enough just to read about something and assume that you'll then be able to do it. That's why in each chapter you'll find lots of questions organized into three different types of sections that are designed to help you test your knowledge:

- **Pop Quizzes:** There are several quizzes per chapter designed to give you a chance to test your understanding of a topic right after reading about it. That way, if you have trouble with the quiz, you know that you should go back and review that section before moving on. Some of the quizzes contain new information or interesting applications that you won't find in the text, so they should be considered "required reading."
- **Algebra Tutorials:** Most chapters contain one or more tutorials designed to teach you a particular algebra technique or walk you through the method for solving a certain type of algebra problem. These tutorials encourage your active participation by occasionally leaving a step unfinished so that you can make the final mental connections and fill in

the details or by prompting you with questions to help you test your understanding.

- **Final Exams:** After you've gone through all the Pop Quizzes and algebra tutorials in each chapter, you'll find one final test of your understanding of the material you just learned in the chapter final exam. Many of the problems in these "exams" contain applications of the material in the chapter that we haven't talked about before. So like the quizzes, these exams should be considered "required reading" . . . and required doing for that matter!

But having quizzes, tutorials, and exams with questions to help you test your understanding is really only useful if you can check whether or not you've answered those questions correctly. So, at the end of the book you'll find a complete and thorough solutions guide containing *full explanations* of how to solve all the questions in the book. Which means that if you're stuck on a quiz or final exam question, you can always turn to Solutions and learn how to get yourself out of that jam. It's like having a personal tutor right in your book.

In addition to these sections that are designed to test your knowledge, you'll find several other reoccurring sections that will help you understand algebra better and make it fun. Amazing, I know.

- **Secret Agent Math-Libs:** Each chapter contains a math-lib section that follows a day in the life of a math secret agent. Yes, you heard that right—these sections do indeed take you on an algebra-inspired spy adventure full of codes to crack and mysteries to unravel. And while you should definitely enjoy the puzzles and twists in these sections, you'll also be dealing with real algebra problems . . . many of which will give your brain a real workout and get you thinking about math in a whole new way.
- **Math Brain Game:** Each chapter also contains a Math Brain Game section designed to get you to think about

algebra problems in new and interesting ways . . . often visually. These sections typically give you some background information, then introduce you to the problem, and finally leave you to think about how to solve it. Of course, this is then followed up with a thorough explanation of how to actually do the solving . . . just in case you didn't figure it all out yourself.

- **Overachiever Badge:** When it comes to algebra, there are a ton of interesting things we could talk about. And there are probably two tons of interesting applications we could talk about. But, obviously, we don't have time to cover three tons worth of stuff. So certain topics that are interesting but not absolutely essential are discussed in the Overachiever Badge sections. And if you read all of these sections, you'll earn an Overachiever Badge! Who doesn't want that?

- **Watch Out!:** Every road has its tricky points that can potentially lead to danger: the curve that came out of nowhere, the giant pothole that opened up overnight, the stop sign that you swear wasn't there yesterday. And the road to algebra is no different. But I'll help you avoid those "dangers," aka common mistakes, in the Watch Out! sections.

- **Algebra Decoder!:** There's a lot of "jargon" in algebra. And by that I mean that there are a lot of algebra-specific words and terms that you don't use in everyday conversation. For many of these terms, you'll find Algebra Decoder! sections that will fill in the details about what they mean and where they come from.

- **Quick and Dirty Tips:** Since this is *The Math Dude's Quick and Dirty Guide to Algebra,* and my podcast is called *The Math Dude's Quick and Dirty Tips,* I'd be cheating you if I didn't give you some quick and dirty tips in the book. Which is exactly what you'll find sprinkled throughout the book.

HOW SHOULD YOU USE THIS BOOK?

There are two options for using this book. The first is to just read it . . . and that's no joke. A lot of math books are filled with far more equations than words and include scant explanations. But this book is different. If you thumb through its pages, you'll see lots of words to describe the topics and algebra problems that we're learning to solve . . . and also lots of drawings to help you visualize their meaning and put a tangible spin on an otherwise abstract subject.

The book was written with the reader in mind, and if you do read it from cover to cover and work through the many quizzes, tutorials, and final exams that it contains, you'll come away with a top-of-the-class deep understanding of algebra.

But I know that not everybody in the real world will have time to read the entire book—and that's okay. Yes, it's perfectly fine to look up and read only the sections that you need help on and skip the rest . . . for now, at least. Of particular use to you if you're this type of reader are the tutorials and quizzes. My advice is to first go through the tutorial on the topic that you're studying, and then take a look at the follow-up practice problems and their solutions to help you figure out how to solve the particular problems you're dealing with.

But I hope, in the end, that no matter which type of reader you are, you'll end up wanting to read more . . . because I genuinely think that if you allow yourself to get into and discover algebra for what it really is, you'll like it. And I genuinely hope that you enjoy reading and working though this book as much as I've enjoyed writing it!

So, here's to hoping we spend a lot more time together learning algebra.

PROLOGUE

Why Math Isn't an Awful Nerd

> Pure mathematics is the world's best game. It is more absorbing than chess, more of a gamble than poker, and lasts longer than Monopoly. It's free. It can be played anywhere—Archimedes did it in a bathtub.
>
> —Richard J. Trudeau, *Dots and Lines*

There's a giant conspiracy in the world to make people think that math is awful. And a lot of people have bought into it. To be honest, I don't blame them—or perhaps you—because, frankly, math can be awful when it's presented to you in a boring and awful way. But I also know that math, and particularly algebra, doesn't have to be that way—that there's another side to it that you've maybe never seen. Math is kind of like that quiet kid in class that you always assumed was a big nerd but was actually rockin' out with her band every weekend on her way to becoming a big-time star. Stick with me for a while and I'll convince you that even though you might think that math is that kid you never wanted to talk to, it's actually that creative, fun, and surprisingly cool rock star.

To give you an idea of what I'm talking about, let's take a minute before diving into the nitty-gritty details of algebra to play a little game of ye olde math. This will probably sound strange, but the objective of this game is to first try and figure out what happens when we add up sequences of positive odd numbers, and then to figure out why it happens! Sounds like fun, right? Oh, not so much? Okay, I totally understand that adding up a bunch of positive odd numbers sounds like a really odd (yes, pun intended) way to spend part of your day. So why would you want to pay any attention to me and play this game?

Because whether or not you've realized it yet, life is all about learning, exploring, and trying new things. In other words, discovering and solving problems you haven't seen before is good for you! Each and every time you do it, you open up a new door that will help you solve other problems that you'll eventually face in your life . . . some of which might even be algebra problems. And believe it or not, it turns out that playing games with math can be a lot of fun too. So hear me out, because I think you're going to end up enjoying this. And if you don't, you can always send a note to mathdude@quickanddirtytips.com and tell me I'm crazy!

BASIC NUMBER PROPERTIES

Since this is a book about algebra, I'm going to assume you're already familiar with the basic properties of numbers—including terms like positive, negative, even, and odd. But since this is the very beginning of the book, let's make sure we're all starting on the same page by quickly summarizing the properties we need to know for our game—namely, the properties of positive odd numbers.

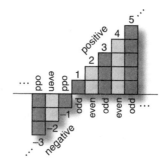

Positive odd numbers are numbers like 1, 3, 5, 7, 9, and so on. Start at 1 and add 2 to get the next one. Add 2 more to get the next one. Do it again, and again, and so on, forever. The positive part means that the number is greater than 0, whereas negative numbers are less than 0. And the odd part means that the number is not evenly divisible by 2, whereas even numbers are always evenly divisible by 2 . . . that's why they're "even"! We'll have more to say about number properties in chapter 3.

POP QUIZ: NAME THAT NUMBER PROPERTY

Use these Pop Quizzes to make sure you understand what we're talking about. If you get stuck or want to check your work, solutions are given at the end of the book.

Again, just to make sure we're all on the same page, here are a couple of quick quiz questions with which to double check yourself. Yes, these should be easy.

	1	99	−5	6	0	24	−8	−1
Am I even (E) or odd (O)?	___	___	___	___	___	___	___	___
Am I positive (P) or negative (N)?	___	___	___	___	___	___	___	___

P.S. There's one "trick" question here. Can you find it?

BASIC ARITHMETIC

I'm also going to assume you're comfortable with the big four basic operations from arithmetic: addition, subtraction, multiplication, and division. That means you should be comfortable with doing things like adding, subtracting, and multiplying a list of numbers, and doing long division. To keep everything as clear as possible, here's a rundown on the notation (in other words, the mathematical symbols) that we're going to use for these all-important arithmetic operations:

- **Addition** will be indicated by a traditional "plus" sign like "+" (no big surprise). So a simple addition problem will look like $1 + 2 = 3$.

- **Subtraction** will be indicated by a traditional "minus" sign like "–" (again, no big surprise). So a simple subtraction problem will look like $3 - 2 = 1$.

- **Multiplication** will be indicated by a dot like "•" (which may be a surprise). So a simple multiplication problem will look like $3 \cdot 2 = 6$. Why do we use a dot instead of the usual "times" symbol as in $3 \times 2 = 6$? Well, let's just say that the letter "*x*" is going to play a big role in our algebraic story as what's called a variable, and that the letter "*x*" looks an awful lot like the "X" symbol traditionally used for multiplication in elementary math. And not to freak you out, but sometimes we'll even omit the "•" entirely! For example, $x \cdot y$ is the same thing as *xy*, $2 \cdot x$ is the same thing as *2x*, and $(1 + 2) \cdot (3 + 4)$ is the same thing as $(1 + 2)(3 + 4)$. It looks like a lot to remember right now, but you'll get used to it quickly.

- **Division** will typically be indicated by a slash symbol like "/" (which again may be a surprise). So a simple division problem will look like $6/3 = 2$. Why not just use the "÷" symbol for division that we used in elementary math? Well, for now, you'll just have to trust me that the traditional "÷" sign won't work very well for a lot of what we'll be doing in algebra. Exactly why will become clearer and clearer as we move through the book.

LET THE GAME BEGIN!

Okay, let's get going adding up those positive odd numbers we talked about. To get started, let's take a look at the following sequence of problems:

1 = 1   ←   This one is *really* easy . . . there's nothing to do!
1 + 3 = ___?   ←   Now they're "tougher" . . . okay, still easy.
1 + 3 + 5 = ___?   ←   *Still* easy? Yeah, pretty easy.
1 + 3 + 5 + 7 = ___?   ←   How about now?
1 + 3 + 5 + 7 + 9 = ___?   ←   And now?

So, what amazing thing do you think will happen when you find all these sums? Will trees start walking and talking? Will the universe implode? Will you get hungry? Well, other than that last one I have serious doubts that any of these things are going to happen. But let's do the experiment anyway since we'll never know unless we try. And maybe . . . just maybe . . . we'll discover something that's mathematically interesting too.

Now, let's make a list of the sequence of answers to the addition problems from before. The first number on our list is 1 . . . since it's the first positive odd number and we're not adding anything to it. To get the second number we need to add together the first *two* positive odd numbers. That's 1 + 3 = 4. Okay, we're on a roll and so far our list looks like

I should point out that the ". . ." symbol just means that there are more numbers yet to come. So, what's the next one? Well, we need to add together the first *three* positive odd numbers: 1 + 3 + 5 = 9. So now our list looks like

Then we add the first *four* positive odd numbers: $1 + 3 + 5 + 7 = 16$, and then the first *five*, and so on. The pattern that emerges looks like

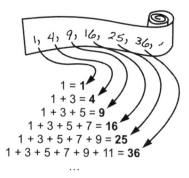

$$1 = \mathbf{1}$$
$$1 + 3 = \mathbf{4}$$
$$1 + 3 + 5 = \mathbf{9}$$
$$1 + 3 + 5 + 7 = \mathbf{16}$$
$$1 + 3 + 5 + 7 + 9 = \mathbf{25}$$
$$1 + 3 + 5 + 7 + 9 + 11 = \mathbf{36}$$
$$\ldots$$

Again, the "..." at the end just shows that we could keep on doing this forever to create a longer and longer list. Although I'll spare you the pain of actually doing that since I'm sure we both have better things to do with our lives. Now, do you see anything interesting in that list? Perhaps there's a pattern developing? If you do see something, that's awesome! If not, don't worry. Just like learning to solve crossword, Sudoku, or KenKen puzzles, it takes a while to develop an eye for this sort of thing. Believe it or not, after a bit of practice, patterns will just start jumping out at you.

- -
SECRET AGENT (in Training) MATH-LIB
- -

IS MATH BORING?

Each chapter includes a "secret agent math-lib" like this one in which you get to solve spy-caliber puzzles using the math we're learning. As if that weren't cool enough, each math-lib is part of a big spy saga that continues throughout the book. Which means that you have to read the whole thing to figure out how the story ends! If you get stuck, hints and solutions are available at the end of the book.

Good evening, agent in training _____ *(name of fruit or vegetable)*. Your mission in life, if you choose to accept it, is to use your perceptive powers of reasoning to solve math puzzles that could save _____ *(name of a planet)* from destruction! I know that's a lot of pressure, so we're going to start things off nice and easy...at least until you graduate from secret agent math academy.

Okay, it's time for your first training session—and it's an important one because we are once and for all answering the question: "Is math boring?" Here's how it's going to work. First, I need you to think of some whole numbers:

(A) Which number between 1 and 10 is written in English using the same number of letters as the number itself? _____

(B) What number between 1 and 10 rhymes with a number between 11 and 20? _____

(C) How many times do you use the numeral "1" when you write all the whole numbers from 10 to 20? _____

(D) Which number between 1 and 10 has the most occurrences of the letter "e" and has not yet been used in this list? _____

If you've got four whole numbers in the spots above, then you've passed your first test. Next, you need to fill in the following blank spaces with the numbers from above (the letters in parentheses show you where to put the corresponding numbers), and then you need to figure out if the answer to each of the following problems is either an even (E) or an odd (O) number. As we'll talk about in more detail later, when part of a problem is written inside parentheses, it means that you need to do that part first. For example, the problem $4 \cdot (2+1)$ means first add $2+1=3$, and then multiply this sum by 4 to get $4 \cdot 3 = 12$. Got it? Okay, here we go:

(1) Is _____ / (_____ + _____) even (E) or odd (O)? _____
        (C)   (A)   (B)

(2) Is _____ − _____ + _____ even (E) or odd (O)?   _____
        (C)   (B)   (A)

Now, you need to figure out if the answer to each of the following problems is either a positive (P) or a negative (N) number. Remember, positive numbers are bigger than zero, and negative numbers are smaller than zero.

(3) Is _____ − (_____ • _____) positive (P) or negative (N)? _____
        (B)   (A)   (D)

(4) Is _____ + _____ − _____ positive (P) or negative (N)?   _____
        (B)   (D)   (A)

Okay, but what about the big question we set out to answer at the beginning of this training session: Is math boring? Well, I'm sure you already know the correct answer ("No!"), but just in case, this should help clear it up for you. All you need to do is insert the letters "E," "O," "P," or "N" from the answers to questions (1) through (4) into the following:

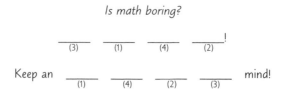

Is math boring?

_____ _____ _____ _____!
 (3)   (1)   (4)   (2)

Keep an   _____ _____ _____ _____ mind!
           (1)   (4)   (2)   (3)

By the way, if you have trouble finishing any of the "Secret Agent Math-Libs" in the book (there's one in each chapter), the answers are available by chapter in Math Dude's Solutions. But no peeking…unless you really need help!

LOOKING FOR PATTERNS IN NUMBERS

Let's now return to the list we made earlier by adding up positive odd numbers. Whether you saw a pattern in this list right off the bat or not, let's walk through the process of finding one. The first thing you might notice is that the numbers in the list 1, 4, 9, 16, 25, 36 keep getting bigger. Is that a pattern? It sure is. But is that an interesting pattern? Well, that's kind of a subjective question, but in this case I'd say not really. After all, the list was created by adding up larger and larger positive numbers, so the sum had to keep getting larger too.

Since I'm so picky about patterns, let's set our sights on finding something a little more insightful. To give us a fresh look at the problem, let's try coming at it from a different angle. Imagine we have several groups of blocks, each of which is connected together to form what we'll call a strand of blocks.

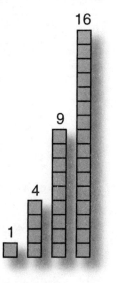

Besides having a solitary block, you're lucky enough to have three other groups of blocks too: a 4-block strand, a 9-block strand, and a 16-block strand. Do those four numbers—1, 4, 9, and 16—look familiar? They should by now—they're the first four numbers from our list. So how is it that these blocks are going to help us find a pattern . . . hopefully even an "insightful" pattern? Well, imagine that all the blocks in each strand are held together by a rubber band that both holds the blocks together but also allows you to move them around and change their position relative to one another. Now, what do you think of doing this?

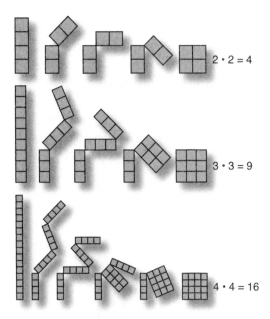

That's pretty interesting, right? The strands of 4, 9, and 16 blocks can be rearranged to form a 2-by-2, 3-by-3, and a 4-by-4 square. Of course, the single block also forms a 1-by-1 square all by itself without doing anything to it. Which means that the first four numbers from our list can all be folded into perfect squares! Not only is that interesting, I'd say it's *incredibly* interesting since most numbers in existence can't be folded into squares like this. And yet every single one of ours can—which means we must have some pretty special numbers on our hands.

EXPONENTS AND PERFECT SQUARES

Multiplying a number times itself is called squaring that number—which seems like a pretty appropriate name given what we've found out about making squares. In other words, "three squared" is just another way to say 3 • 3. But a big part of math is about coming up with better and more efficient ways to write things, so people have developed another bit of notation to describe squaring numbers that looks like this:

$$1 • 1 = 1^2 = 1$$
$$2 • 2 = 2^2 = 4$$
$$3 • 3 = 3^2 = 9$$
$$4 • 4 = 4^2 = 16$$

The superscript number is called the "exponent"—in this case it's 2—and it tells you how many times to multiply together the "base"—that is, the number right before the exponent. So,

$$2 • 2 = 2^2 = 4$$
$$2 • 2 • 2 = 2^3 = 8$$
$$2 • 2 • 2 • 2 = 2^4 = 16$$

and so on. If you continued this pattern and kept adding more and more lines, you'd be finding what are called higher and higher "powers of 2" . . . that is, 2^5, 2^6, 2^7, and so on.

On a related note, numbers formed by squaring whole numbers (those are numbers that don't have fractional parts) are known as "perfect squares." That means that 1, 4, 9, and 16 are all perfect squares since $1^2 = 1$, $2^2 = 4$, $3^2 = 9$, and $4^2 = 16$. And there are a lot more perfect squares too . . . an infinite number, in fact. Just pick any whole number you like, multiply it by itself, and you've found yet another perfect square. You may have noticed that 1, 4, 9, and 16 are also the numbers that we've written on our list so far . . . and they're all perfect squares! If you're looking for more information on exponents right now (and you just can't wait), you'll want to check out chapter 4.

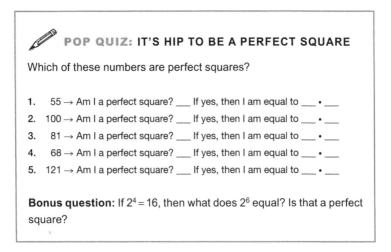

A SURPRISING SEQUENCE OF NUMBERS

Okay, let's get back to looking at that pattern of perfect squares we've discovered. The first thing we should think about is whether or not the pattern keeps on going forever. In other words, will every number we ever write on our list be the square of some other number? So far, we've verified that 1, 4, 9, and 16 are all perfect squares. Let's now look at a few more numbers and see if the pattern holds up. The next number on the list is 25, and 25 is equal to $5 \cdot 5 = 5^2$.

So, yes, that one works too. And the number after that is 36, which is equal to $6 \cdot 6 = 6^2$. Again, that one works. And if we keep going, we'll find that no matter how big a number we get to, the pattern will continue! For example, here's the pattern written all the way out to the tenth number on the list:

1) $1 = \mathbf{1} = 1^2$
2) $1 + 3 = \mathbf{4} = 2^2$

3) $1+3+5=\mathbf{9}=3^2$
4) $1+3+5+7=\mathbf{16}=4^2$
5) $1+3+5+7+9=\mathbf{25}=5^2$
6) $1+3+5+7+9+11=\mathbf{36}=6^2$
7) $1+3+5+7+9+11+13=\mathbf{49}=7^2$
8) $1+3+5+7+9+11+13+15=\mathbf{64}=8^2$
9) $1+3+5+7+9+11+13+15+17=\mathbf{81}=9^2$
10) $1+3+5+7+9+11+13+15+17+19=\mathbf{100}=10^2$

Take a look at this pattern of numbers. Not only is each sum the square of some number, but each sum is actually the square of the number that gives its order in the list. In other words, to find the value of what we'll call the "nth" number on our list (we're just using the letter "n" to represent whichever number on the list we're interested in at the time . . . it could be 1, 2, 3, 4, or whatever), all you have to do is calculate the value n^2. For example, the first number on the list is represented by $n=1$, and therefore has a value of $1^2=1$; the second number on the list is given by $n=2$ and has a value of $2^2=4$; and so on until we get to the tenth number on the list, represented by $n=10$, which must have a value of $10^2=100$. And although we haven't written them, we could go on and on to as high a value of n as we'd like.

Okay, so that's all pretty cool, but I've got an even more impressive way to look at the numbers on our list. Check this out:

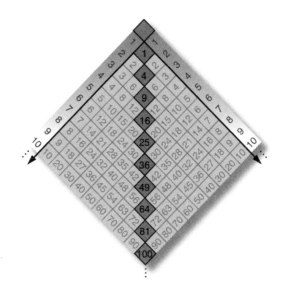

It kind of looks like an awesome game board, right? Well, it sort of is . . . because it's your dear old friend the multiplication table. But notice the numbers running from top to bottom right down the middle of the table. Those numbers are all the perfect squares. In other words, those are all the numbers on our list! Each and every one of them, somewhat magically, turns out to fall right smack-dab on the line that goes down the middle of the multiplication table.

RECAP

Are you starting to agree with me that our list of numbers is pretty amazing . . . or at least that it's pretty surprising? To really see why I'm so enthusiastic about it, let's take a step back and think about what we've actually done so far. We started by adding up a bunch of odd numbers. Not a big deal . . . yet. Because it turned out that when we did this several times in a row, we ended up creating a list of very special numbers known as perfect squares—1, 4, 9, 16, 25, 36, and so on. And, if you think about it for a minute, this seems pretty strange. After all, why should adding up odd numbers have anything to do with creating perfect squares? They're *totally* different beasts. And we're not talking horses versus ponies different—we're talking mosquitoes versus unicorns!

So, have we discovered an ancient secret soon to bring us all fame, fortune, and glory? Well, sorry to disappoint you but although our discovery is awesome, it isn't going to make us rich. Bummer, I know. But there is a silver lining here because although the solution may not directly bring you wealth, it will bring you something equally as important: knowledge (which, of course, could eventually turn into money). I know, I know, you'd rather have the cash . . . sorry! But let's not lose track of what we're really doing here—that is, trying to understand how and why the puzzle we've looked at works. Believe it or not, the solution is surprisingly simple. But don't be fooled by the word "simple" here, because the solution is also extremely elegant and, to my eyes at least, beautiful. So, let's figure out how it works.

<pre>
  ┌──┬──┬──┬──┬──┬──┬──┬──┬──┬──┬──┬──┬──┬──┬──┬──┬──┐
  │ │ │ M│ A│ T│ H│ │ B│ R│ A│ I│ N│ │ G│ A│ M│ E│ │ │
  └──┴──┴──┴──┴──┴──┴──┴──┴──┴──┴──┴──┴──┴──┴──┴──┴──┘
</pre>

A POSITIVELY ODD PUZZLE

Each chapter includes a "math brain game" section like this one that's designed to put your brain into overdrive and get you thinking about algebra in new and unexpected ways. And they're usually pretty fun too!

STEP 1—THE GAME PLAN

Okay, we're going to solve our positively odd puzzle using strands of blocks . . . exactly like the ones we "folded" into perfect squares earlier. Let's use five strands: a single block and four made up of groups of 3, 5, 7, and 9 connected blocks. Do you recognize those numbers? We've seen them enough at this point that I bet you're starting to get tired of them. Of course, they're just the first five positive odd numbers. Believe it or not, using nothing more than these five strands of blocks, we can show why the incredible relationship we've discovered works. So, your job is to figure out how to arrange the strands of blocks on the grid below to prove that the sums of sequences of positive odd numbers, such as $1+3$, $1+3+5$, $1+3+5+7$, and so on, must give numbers that are equal to perfect squares.

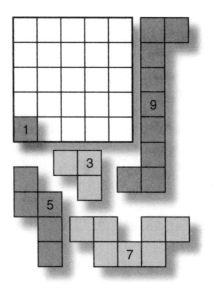

Here's a hint: The first block is in a good spot . . . so don't move it! And remember that besides placing the connected strands on the grid, you can also rearrange their shapes like we did before when we first discovered that the strands of 4, 9, and 16 blocks could be "folded" into perfect squares. So, give it a shot, and then read on when you either figure it out or get stuck and need some help!

STEP 2—CAN YOU TAKE A HINT?

Okay, I'm now going to put the strand of blocks for the next number on our list, 3, onto the grid. And to help out a little more, I'm also going to rearrange the shapes of the remaining 5, 7, and 9 block strands. Does that help you see the solution to the puzzle?

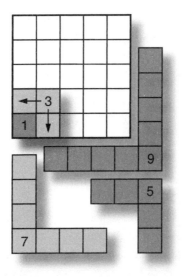

Remember, the goal is to come up with a way to show that whenever you start with 1 and add any number of the subsequent sequence of odd numbers to it, you get a number that's a perfect square. If you think about it, you'll see that this means that all the puzzle pieces (that is, all the strands of blocks) need to combine to form perfect squares within the large square grid (since the grid is a square!). Do you see how after we put the 1-by-1 and 2-by-2 squares on the board, the resulting shape is a perfect square? Can you use the remaining shapes to make 3-by-3, 4-by-4, and 5-by-5 perfect squares too? At this point, I'm close to giving the answer away . . . so I'll stop talking for now. In the meantime, think about it some more and then move on when you're ready for the answer.

STEP 3—THE SOLUTION!

Okay, it's the big moment we've all been waiting for—here's the solution to our positively odd puzzle:

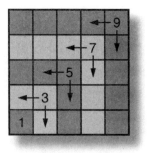

It really is a thing of beauty, right? This drawing shows that adding up a sequence of odd numbers and getting perfect squares is,
somewhat surprisingly, not a mystery . . . not at all, in fact. It's simply a result of the way that shapes fit together. That is, each successive odd number perfectly "wraps around" two sides of the square
formed by the sum of the previous odd numbers. For example, the
three blocks from the 3-block strand perfectly wrap around the single block, the five blocks from the 5-block strand perfectly wrap
around the three blocks from the 3-block strand, and so on. Do you
see why I've called it an elegant solution?

WRAP-UP

And with that we've arrived at a critical junction in our
relationship—the point at which you decide whether or not to
stick around and learn algebra with me. Some of you will see the
elegance in the solution to our positively odd puzzle. Some will
not. And that's okay! After all, we all have different tastes and
preferences about food, cars, clothes, art, movies, music, and yes
even math. For some, seeing a solution to a puzzle like the one
we've looked at is genuinely exciting. Others are left wondering
about the sanity of those who are now excited. But just because
you're in the skeptical camp right now doesn't mean you won't
learn to love math . . . at least a bit. I should know since, once
upon a time, I was one of you! Honestly, I didn't always love

math. But once I was introduced to "real" math—the kind that can be used to discover and explain things—I was hooked.

So, really, what's the point of all this? This puzzle, this prologue, this book? Well, first of all, we're going to learn algebra, and we're going to learn it well. And when I say learn, I'm not talking about memorizing sequences of steps and formulas for the sole purpose of acing your next test. No, I'm talking about learning algebra so that you really *understand* it. And that will not only help you ace your next test—with flying colors—but it will also help you ace whatever other challenges you face in your life. But that's still not enough for me! I'm not satisfied with just helping you "do" and understand algebra. I want to show you the bigger picture too—the picture that makes some people look at the puzzle we solved with a sense of delight. It may sound ambitious, but in addition to teaching you algebra, I want to help you discover the world's oldest, cheapest, most practical, powerful, fun, and undeniably amazing game.

Hello, world! It's me, math.

· FINAL EXAM ·

Each chapter ends with a summary of the main topics covered in that chapter, and a series of "exam" questions about those topics. This gives you a chance to review the main ideas from the chapter and make sure you understand them. If you don't, go back and review the topic before moving on . . . or else you're begging for confusion! If you get stuck, or you want to check your work, answers are available in Math Dude's Solutions.

• NUMBER PROPERTIES

1. Think of an even number: _____.
   Now think of another even number: _____.
   - Add them together: _____.
     Is the answer even or odd? _____.

- Subtract the first from the second: _____.
  Is the answer positive or negative? _____.
  Is the answer even or odd? _____.

- Multiply them together _____.
  Is the answer even or odd? _____.

2. Think of an odd number: _____.
   Now think of another odd number: _____.

   - Add them together: _____.
     Is the answer even or odd? _____.

   - Subtract the first from the second: _____.
     Is the answer positive or negative? _____.
     Is the answer even or odd? _____.

   - Multiply them together _____.
     Is the answer even or odd? _____.

3. Think of an even number: _____.
   Now think of an odd number: _____.

   - Add them together: _____.
     Is the answer even or odd? _____.

   - Subtract the first from the second: _____.
     Is the answer positive or negative? _____.
     Is the answer even or odd? _____.

   - Multiply them together _____.
     Is the answer even or odd? _____.

Do you see any patterns in these solutions? Perhaps something to do with combinations of pairs of even and odd numbers?

- **NOTATION FOR ARITHMETIC**

Translate the following sentences into arithmetic problems. And be sure to use our fancy algebra-ready notation for addition, subtraction, multiplication, and division.

4. "Thirty-two plus twenty-eight minus fifty-nine." ....

   ____ 32 + 28 − 59 ____

5. "Eighteen times two divided by nine times two." ....

   _____

6. "Eighteen times two, all divided by nine times two." ....

   _____

• SQUARING NUMBERS

7. Pick a number between 1 and 10: _____. Let's call this number a.

   - Square that number → $a^2 =$ _____. Let's call this number b.

   - Multiply your original number by 2 → $2 \cdot a =$ _____. Let's call this number c.

   - Subtract c from b → $b - c =$ _____. Let's call this number d.

   - Add 1 to d → $d + 1 =$ _____. Let's call this number e.

8. Now start again with the same number, a, that you picked before.

   - Subtract 1 from that number → $a - 1 =$ _____. Let's call this number f.

   - Square this number → $f^2 =$ _____. Let's call this number g.

The numbers e and g are the same, right? Surprise! Want to know why? Good, I'm glad. Here's a hint: It's not magic, it's algebra. And we'll find out exactly why it works by the time we get to the end of the book. Sorry, there's no shortcut to learning.

• SEARCHING FOR PATTERNS

Can you find the patterns in the following lists of numbers. In other words, can you find the key thing that changes from number to number in each list that allows you to predict what the next number in the list will be? Some of these are pretty tough, so don't be discouraged if you can't figure them all out. It's okay to peek at the solutions.

9. 1, 2, 4, 8, 16, 32, . . . What's the next number? _____64_____

10. 2, 5, 8, 11, 14, 17, . . . What's the next number? _____20_____

11. 1, 3, 7, 15, 31, 63, . . . What's the next number? _____127_____

12. 1, 1, 2, 3, 5, 8, 13, . . . What's the next number? _____21_____

PART I

WHAT IS ALGEBRA, REALLY?

_____

We learn the essentials of algebra and find out how to "take it to the streets" to help us get stuff done.

Though this be madness, yet there is method in't.

—William Shakespeare

CHAPTER 1

Taking Algebra to the Streets

> If people do not believe that mathematics is simple, it is only because they do not realize how complicated life is.
>
> —John von Neumann

In case you haven't heard the news, it turns out that life is not just about solving puzzles, playing games, and having fun. It's too bad, I know. But nevertheless it's true—sometimes we simply need to get stuff done. And it's in those situations that algebra really shines. Why? Well, think of algebra as the highway and road system of the math world. Just as it's tough to drive your car anywhere without roads, it's also tough to solve most math problems without some algebra. But, you might be wondering, I've been solving math problems since I was a little kid, and I don't ever remember using algebra before! What gives? Well, I've got some news for you . . . and it may come as a surprise: without even knowing it, you have slowly but surely become something of an algebra expert.

In this chapter you'll discover all the algebra you've already been doing, which should help you realize why algebra is so useful and important (yes, really). I'll also introduce you to some math concepts that are crucial to understanding algebra: things like variables, inequalities, square roots, graphs, and more.

THE SECRET ALGEBRA YOU'VE ALREADY BEEN DOING

To give you a taste of what I'm talking about, let's take a look at a specific example from your everyday life. The first thing I want you to do is put down this book and go get your wallet. Really, go ahead and get it—I don't mind the wait. You back? Great. Now, all you need to do is simply count how much money you have. So go ahead, dive in there and count up that cash.

Okay, do you have your number? Once you do, feel free to safely stow your wallet—we won't be needing it anymore. Just be sure to keep that number in your head.

Why do we need to know that? Well, here's the scenario we're dealing with: You and I are planning to have some friends over tonight for dinner. Our plan is to barbecue a pizza, but we've been so busy lately that we completely forgot to shop for party food. Oops! We've only got an hour to go before our guests arrive, and the refrigerator is a total blank canvas. In other words, it's a cold and cavernous empty space . . . which means that we need to buy everything: pizza dough, tomato sauce, cheese, pepperoni, and even charcoal. Which brings us to the point of me having you check what's in your wallet: I'm counting on you to pick up the groceries. So, do you have enough money to buy them?

Presuming you're older than, let's say, ten and that you've done a little shopping in your time, I imagine you could figure this out without any problem. In fact, I'd wager that you stumble across little quandaries like this almost every day, and you rarely—if ever—think twice about them. But what are you actually doing when you solve them? Well, hold on tight for a few minutes, because we're about to go through this in what some might call "excruciating" detail. But, rest assured, there's a payoff at the end—I promise. Okay, here we go. Whether or not you've realized it before, there are basically three steps for answering a question like this:

STEP 1

First, you need to draw upon your past experiences to estimate how much each item on your shopping list will cost. There's no need to be super precise here, we're just looking for a ballpark figure—say, to the nearest dollar. In the case of our party food, I'd guess the charcoal will cost about $5, the cheese and pepperoni will cost about $3 each, and the dough and tomato sauce will cost somewhere around $2 apiece.

STEP 2

Next, you need to add these individual amounts together to estimate the total cost of the shopping trip. In this case it's pretty simple:

$$\$5 + \$3 + \$3 + \$2 + \$2 = \$15.$$

Which means that you think your trip will cost around $15. Remember, we're just making a rough estimate of the cost here, so we could easily be off by a few bucks either way.

STEP 3

The final step to figuring out whether or not you need to go to the bank before you head to the grocery store is to compare your total estimated cost—which is $15—to the amount of money you have in your wallet (you do remember how much you have in your wallet, right?). So, if you found $10 in your wallet earlier, then since $10 is less than $15, you know that you'd better go to the bank or else you run the risk of coming up short at the grocery store. On the other hand, if you have $20 in your wallet right now, then since $20 is more than $15, you should have plenty of money for the groceries—even if we were a dollar or two off with our estimate. So, if you and I were actually planning a party, where would you be headed right now? To the grocery store or to the bank?

> *"Wow, thanks! All of that effort just to solve one little problem that I, uh, already knew how to solve anyway. Great job, Math Dude."*
>
> —Anonymous Reader

Ouch! The truth can hurt. But, nonetheless, this anonymous reader is correct: taken at face value the (over-)analysis of this problem was over the top and overly complicated. But, as I said earlier, there's a lot more to this problem than meets the eye, and the payoff is coming. Well, actually, it's here.

The truth is that the mathematical part of this problem had nothing to do with pizza or the contents of your wallet. Yes, in practice we were adding up numbers representing quantities of money and seeing if we had sufficient funds to buy the things on our grocery list. But mathematically we were doing much more—we were estimating unknown values, assigning them to variables, performing integer arithmetic, and solving greater than or less than inequalities. In other words, we were doing the stuff of real math. In fact, we were doing the stuff of real algebra!

Does that surprise you? Perhaps it does, but hopefully it's a good surprise since you now see that math really does apply to many things you do every single day of your life. And, perhaps more important, hopefully you now see that you already know some algebra. Certainly not all of it, but some of it—and that's a great start.

INEQUALITIES: <, >, ≤, AND ≥

I mentioned that one piece of the algebra you were unknowingly using before was the idea of an inequality. I bet you've had a run-in or two with less than and greater than symbols at some point in your life, but a little refresher on the topic won't hurt anybody. Inequality symbols are so named because they, unlike an equals sign, tell us when two things are *not* equal. For example, here's how the less than ("<") and greater than (">") symbols work:

- $3 < 4 \rightarrow 3$ is less than 4 .... Yes! ✓
- $3 > 4 \rightarrow 3$ is greater than 4 .... No! ✕
- $4 > 3 \rightarrow 4$ is greater than 3 .... Yes! ✓
- $4 < 3 \rightarrow 4$ is less than 3 .... No! ✕

And while we're at it, we may as well talk about two additional inequality symbols that we'll run into in the next chapter. The first is called the less than or equal to sign ("≤"), and it works like this:

- $3 \leq 4 \rightarrow 3$ is less than or equal to 4 .... Yes! ✓
- $4 \leq 4 \rightarrow 4$ is less than or equal to 4 .... Yes! ✓
- $5 \leq 4 \rightarrow 5$ is less than or equal to 4 .... No! ✕

◀ Q&DT

The second is called the greater than or equal to sign ("$\geq$"), and it works like this:

- $3 \geq 4 \rightarrow 3$ is greater than or equal to 4 .... No! ✗
- $4 \geq 4 \rightarrow 4$ is greater than or equal to 4 .... Yes! ✓
- $5 \geq 4 \rightarrow 5$ is greater than or equal to 4 .... Yes! ✓

These four symbols give us a quick and convenient way to compare the sizes of two numbers.

POP QUIZ:
WHICH IS BIGGER? SMALLER? EQUAL?

Let's make sure you've got this whole inequality business down pat. Your job is to look at the two things being compared, and then to fill in the symbol that makes the statement true. Each of these five symbols should each be used once: >, <, $\geq$, $\leq$, and =. If you don't know the answer to a question, the Internet is your friend—look it up!

1. The size of the moon is ____ the size of a mountain
2. The numbers 5, 8, 13, and 21 are ____ the number 5.
3. 100 ____ 10^2
4. The size of the planet Saturn is ____ the size of the planet Jupiter.
5. The numbers 5, 8, 13, and 21 are ____ the number 21.

USING VARIABLES

But how exactly was everything we were doing earlier to solve our grocery shopping dilemma algebra? I mean, just saying that it was doesn't actually make it so—we need some concrete details! Fair enough. Let's go over the problem again, but this time let's look at it a bit more algebraically. First, let's think about what to call the amount of money you found in your wallet. Of course we could call it $10, or whatever amount you found. But what if you want to write out a slightly more general "recipe" (no, not the kind for food . . . the kind that tells you how to do something) that you can use over and over again each week to figure out if you'll be able to pay for the things on that week's shopping list. Of course, you don't know how much money you're going to have in your wallet next week, or the week after that, but you'd still like to be able to write out your process for figuring out whether or not you'll need to go to the bank that week.

The bottom line is that it's often just plain useful to represent a number without actually knowing what the value of that number is . . . at least not yet. So, just to get a taste of how this works, let's represent the amount of money you found in your wallet using the letter m. That's right, we're going to use a symbol to represent a number. In algebra (and all of math), we call symbols like m, variables (we'll have a lot more to say about variables in the next chapter). As we've said, we don't actually care what m stands for yet—it could be ten cents, ten dollars, or ten thousand dollars. It doesn't matter. We just need to know that there is *some* amount of money in your wallet. And once we know that, we call it m so that we can talk about it and, as you'll see, use it to solve problems.

Okay, let's go a step further now and figure out a way to describe the process of totaling up the costs of all the groceries you need to buy. But, again, let's not bother with being specific about which items you need to buy right now and how much they're all going to cost—maybe we don't actually know yet! Instead, let's represent the costs of the various items using the symbol c. And let's also add a little number to each c to show which item we're talking about on our list. In other words, c_1 represents the cost of the first item on the list (maybe it's pizza dough), c_2 represents the cost of the second item on the list (perhaps cheese), and so on.

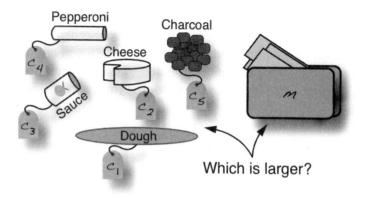

WRITING EQUATIONS

Now, here comes the good part. We can use all these variables that we've defined to write our problem algebraically. For example, you can answer the question "Do you need to go to the bank before the grocery store?" using this

$$c_1 + c_2 + c_3 + \ldots > m$$

How does that work? Well, in words, this says that if you add up all of the costs on your list (those are all the c values), then if that number is greater than the amount of money in your wallet (which we've called m), the answer is "Yes, you need to go to the bank."

ALGEBRA MAKES THINGS EASIER!

But what's the point of all this complicated notation? The point is that it doesn't matter how many costs you have, or what the actual amount of those costs are, or even how much money you have in your wallet, because the algebra we've come up with can deal with it. Not only can it help you avert a foodless party catastrophe, but it can just as easily be extended to figure out if you can afford something like your normal weekly shopping trip or an upcoming vacation—all you have to do is add more c variables for the additional costs.

It's okay if this way of thinking seems a little abstract right now . . . it is! That's exactly the point and the power of it all. But even though algebra may seem confusing at times, the whole point of it is that it helps us solve problems and makes life easier. I know that might be hard to believe when you're struggling with your homework—it certainly isn't making *your* life easier—but rest assured that it'll start making more and more sense as we see more and more examples of algebra in action throughout this book.

Finally, I know that some of this stuff might look a little intimidating. But don't let it scare you off. As we just saw, all of this algebra is really just another way of writing down and dealing with something that's completely familiar to you. In other words, with algebra, things are really no more complicated now than they were before; all that's changed is that we have some new fancy ways to write things. So, don't worry if this all looks like a bunch of gobbledygook at this point . . . we'll be covering all of these ideas again in a lot more detail.

POP QUIZ:
CAN YOU USE ALGEBRA TO MAKE A BUDGET?

You need to buy certain things every month, but you've only got so much to spend. Can you use algebra to help you figure out whether or not you're going to be able to afford what you need? Start by creating variables to represent your monthly expenses. Then create another variable to represent how much money you make each month—either from a job or from your parents. Finally, write down an inequality like the one we wrote to determine if we needed to go to the bank . . . but this time, make the inequality tell you whether or not you can afford what you need to buy! If you need help, check out the solution in Math Dude's Solutions.

Bonus question: Once you answer this quiz question, go back and think about why we used ">" and not "≥" in our earlier grocery shopping problem.

WHAT'S THE POINT OF ALGEBRA?

Okay, we've now been introduced to two decidedly distinct ways to think about math. The first was to picture math as a playful and fun-loving puzzle . . . a game for your brain, as it were. And the second was to think of math as a useful, practical, and down-to-earth tool. So which is right? Of course, they both are. There are a lot of things in the world that are both useful *and* fun: cars and computers, for example. All of which might leave you wondering if these two views ever run into each other? Or do they just quietly live out their separate lives?

Well, to find the answer to this question, let's jump in our algebraic time machine (you'd definitely need algebra to build one of those!) and travel back in time about 4,500 years—back to the time when the Great Pyramid of Giza was being built in Egypt. Imagine, if you will (since no, this isn't a true story), that at that time there was a young dude with a strange hobby: he loved to tie knots in rope. Not surprisingly, everybody else thought this guy (we'll call him "Knot Dude") was borderline bizarre, but he really didn't care because he was absolutely fascinated by his knots.

Now, it turns out that Knot Dude's dad (we'll call him "Papa Knot") was a rather prominent member of Egyptian society. In fact, Papa Knot was the person in charge of figuring out how to build the foundation for the Great Pyramid. Not a job to be taken lightly since this is a *big* pyramid—about 750 feet long on each side, and 450 feet high! Not having much of a social life, Knot Dude spent a lot of time in the evenings listening to his father complain about the difficulty of his job. In particular, one problem was proving quite vexing: how to line up the four walls of the foundation of the pyramid so they all would meet to form a giant square. After all, without a perfectly square base, the pyramid wouldn't be a pyramid—it'd just be a mess.

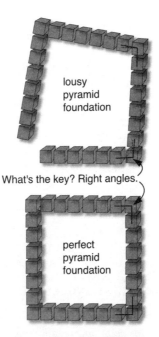

lousy
pyramid
foundation

What's the key? Right angles.

perfect
pyramid
foundation

What's the key to getting the foundation to come out right? Well, Papa Knot knew that if he could come up with a way to make sure that each of the four walls came together and formed perfect corners known as "right angles," then his foundation would also be perfect. But for the life of him, he couldn't figure out how to do it!

ALGEBRA DECODER! 🔍

• RIGHT ANGLE

Extra details about the origin and meaning of some math terms—aka "jargon"—will be given in "Algebra Decoder!" sections like this.

Why is it that we call this type of "square" intersection between two lines a **right angle**? Well, the quick and dirty explanation is

Cont'd

Cont'd from page 43

found by looking at the similarity between the phrase "right angle" and the word "rectangle" (think rect-angle). They almost look and sound the same, right? And this makes perfect sense because all four corners of a rectangle form right angles! On a related note, the symbol used to indicate that an angle is a right angle looks kind of like an "L" (which you can see on the pyramid foundation drawing). Again, this symbol makes sense since it looks like one corner of a rectangle.

POP QUIZ:
HOW MANY RIGHT ANGLES CAN YOU FIND?

Write the number of right angles in each shape on the line. Only count right angles on the *inside* of each shape. In other words, we're not talking about the total number of corners in each shape! This distinction really only matters for one of these shapes—which is it?

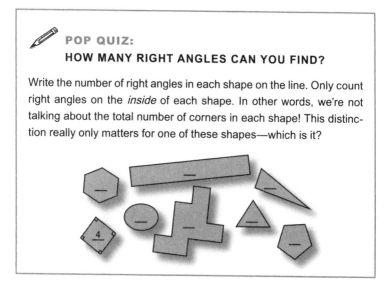

When Knot Dude heard his father describe his problem with building the pyramid foundation, he immediately got excited because he realized that he had made a discovery several months earlier during one of his knot-tying frenzies that solved the problem. All Papa Knot needed to do was get a single piece of rope and tie thirteen evenly spaced knots in it—one at each end, and then eleven more spaced evenly between the two ends. And that's it . . . that's all he needed!

Well, actually, Papa Knot would need to do a little more than just tie the knots—he'd need to use them in a rather clever way too. The trick was to lay out the rope in a straight line, and then fold one end up after three even segments, and the other end up after five even segments. The result was a triangle. But it wasn't just any old triangle, it was a *very* special triangle. In fact, Knot Dude found that as long as the rope was pulled nice and taut, two sides of this triangle will always come together to form a perfect right angle.

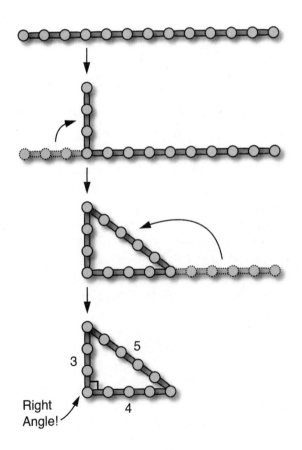

Ding-ding-ding! That's precisely what Papa Knot needed—a right angle that he could use to set the four corners of the pyramid's foundation and ensure that it formed a perfect square. Of course, after Knot Dude told his father, great celebration ensued and a merry time was had by all since the great dilemma had been resolved . . . with math, I might add.

SQUARE ROOTS

Remember when we found all those perfect squares—numbers formed by squaring whole numbers like 1, 4, 9, 16, and so on? Those were good times, right? Yeah, well, the good times are back because I've got a question that's related to perfect squares: If I give you a perfect square, can you figure out what number was squared to get that perfect square? For example, let's think about the perfect square 49. The fact that 49 is a perfect square means that there must be some unknown number that when multiplied by itself gives 49. So, how can you figure out what that number is?

As you may know, we can find that number by taking what's called the square root of the perfect square 49 . . . which we usually write $\sqrt{49}$. And, since we already found that $7^2 = 49$ when we looked at perfect squares, we can immediately figure out that $\sqrt{49} = 7$. Okay, that makes sense, but isn't it also true that $(-7)^2 = 49$? So shouldn't I really have said that $\sqrt{49}$ is equal to either 7 *or* –7. Aren't both of these correct solutions? Well, not exactly. Here's what I mean: In algebra, you'll sometimes see problems that look like

$$x^2 = 49$$

And, as we'll look at in detail in the next chapter, the solution to this problem is either $x = \sqrt{49} = 7$ or $x = -\sqrt{49} = -7$ (you should try plugging both these values of x into the equation to check that they both work). But saying that this equation has **◀Q&DT** two possible numbers that make it true isn't the same thing as saying that $\sqrt{49}$ itself is equal to either 7 or –7. The bottom line is that $\sqrt{49}$ is just a number . . . and it's a positive number!

So, that means that $\sqrt{49} = 7$. That was pretty easy to figure out, right? Perhaps you're thinking that square roots aren't so tough to calculate after all. Well, yes . . . it's not too difficult to find the square root of a perfect square. But don't get too excited because life isn't always so simple. In particular, it's not nearly as easy to find the square root of non-perfect squares. For example, what's the square root of 60? That's a lot harder to figure out because there is no whole number that you can multiply by itself to get 60. So what can we do?

HOW TO CALCULATE SQUARE ROOTS

Exactly how you go about calculating the square root of a non-perfect square number depends upon how accurately you need to know the answer. If you just need to know roughly what **◀Q&DT** something like $\sqrt{60}$ is equal to, then you can come up with a very rough estimate simply by figuring out which two perfect squares 60 falls between. In other words, since $7^2 = 49$ and $8^2 = 64$, we know that $\sqrt{60}$ must be between 7 and 8. And we could also guess that the real answer is a little closer to 8 than 7, since 60 is closer to 64 than 49. Make sense?

Of course, this works for much larger numbers too. For example, what's the approximate square root of 10,400? Well, we know (or at least we can quickly figure out) that $100^2 = 10,000$—which is already pretty close to 10,400. But it's not quite enough. So what's 101^2? If you do the multiplication, you'll find that it's 10,201—closer yet, but still not close enough. Okay, how about 102^2? Well, that equals 10,404—a little bit too much! But nonetheless, 102^2 is very close to 10,400 . . . so we can conclude that $\sqrt{10,400}$ is equal to a tiny bit less than 102.

But what if an estimate just isn't good enough? What if we

Q&DT ▶

need to know the answer with a high degree of accuracy? In that case, your best option is to use a calculator or computer. Sure, there are methods that you can use to help you calculate better and better approximations by hand. But, in all honesty, those methods were developed hundreds of years ago . . . when there weren't machines that could do the job. But now there are . . . and those machines are much faster and more accurate than you are. So doesn't it make sense to use them? After all, we've got better things to do with our brains!

WATCH OUT!

The Peril of Square Roots

There are some things in life . . . and math . . . that you really want to avoid. And those are exactly the kinds of things you'll find in Watch Out! sections.

Before you run off and start calculating square roots, I should warn you that not every number has one! Which numbers are out of the club? Well, let's think about it. When you square a positive number, the answer is always positive. And when you square a negative number, the answer is also always positive. Which means that the square of *any* number is always positive. And, if you think about it, you'll see that this means that a negative number does not have a square root—since there's no number that you can square to equal it (at least none that you know of at this point). So, consider yourself warned—you can't take the square root of a negative number!

> ### 🖉 POP QUIZ: CALCULATING SQUARE ROOTS
>
> Now, it's time for a test of your root-taking talent. For the following problems, I want you to solve one-quarter of them exactly, another one-quarter of them by estimating the square root, another one-quarter of them using a calculator, and, finally, don't solve one-quarter of them at all! You can choose which method you want to use for each problem. But choose wisely . . .
>
> 1. $\sqrt{16} =$ ___   2. $x^2 = 1001$. So, $x \approx$ ___
> 3. $\sqrt{-1} =$ ___   4. $x^2 = 81$. So, $x =$ ___
> 5. $\sqrt{3,14159} \approx$ ___   6. $x^2 = -42$. So, $x =$ ___
> 7. $\sqrt{55} \approx$ ___   8. $x^2 = 111$. So, $x \approx$ ___
>
> Hold on a minute! What do those squiggly equals signs mean? Those little guys indicate that the thing on the left is *approximately* . . . but not *precisely* . . . equal to the thing on the right. In other words, $\pi \approx 3.14$ says that the number π (pi!) is approximately equal to 3.14, but not exactly. Does that help you figure out which one-quarter of the problems you should answer exactly?

THE PYTHAGOREAN THEOREM

Almost everybody has heard of the Pythagorean theorem. Heck, even the brainless Scarecrow from *The Wizard of Oz* knew about the Pythagorean theorem! Remember? When he receives his diploma from the Wizard, he declares: "The sum of the square roots of any two sides of an isosceles triangle is equal to the square root of the remaining side. Oh, joy . . . rapture! I've got a brain!" Well, the only problem is . . . the Scarecrow got it wrong! So, what does the Pythagorean theorem actually say?

In words, the Pythagorean theorem says that the lengths of the two legs of *any* right triangle have a very special relationship to the length of the long side of the triangle (which is called its hypotenuse). Specifically, if you square the lengths of the two legs and then add the resulting numbers together, that number will always equal the square of the length of the hypotenuse. Okay, words are nice, but this is an algebra book, so let's express the Pythagorean theorem algebraically. Here it is:

$$a^2 + b^2 = c^2$$

What do those symbols a, b, and c mean? Well, a and b represent the length of each leg of the triangle, and c represents the length of the hypotenuse. Think about it for a minute, and you'll see that this formula "says" the exact same thing as that other more wordy description—but it's a lot more concise.

The beauty of the algebraic form of the Pythagorean theorem is that I can give you the lengths of the legs of a right triangle (a and b), and you can tell me how long the hypotenuse must be (in other words, c). Just to make sure I'm not crazy, let's check and see if this works for Knot Dude's rope. Remember, Knot Dude's rope triangle had $a = 3$ and $b = 4$. That gives us $a^2 = 9$ and $b^2 = 16$, which means that

$$a^2 + b^2 = 9 + 16 = 25$$

Now, how about c^2? Well, in Knot Dude's triangle $c = 5$, which means that $c^2 = 25$. And, of course, that's exactly the same answer that we got for $a^2 + b^2$. So, who would've guessed it: the Pythagorean theorem actually works! (For the record, I might have guessed it.)

ALGEBRA DECODER!

• WHAT IS A "PYTHAGOREAN"?

Just in case you're wondering where the word "Pythagorean" in "Pythagorean theorem" comes from . . . well, here's your answer: Pythagoras was a Greek dude who lived a *really* long time ago . . . about 2,500 years ago, in fact! Given this length of time, it shouldn't be too surprising that many of the details of Pythagoras' life are a little fuzzy. But we do know that he was a philosopher and a mathematician, and that he started a kind of religious movement known as "Pythagoreanism," where, among other things, everybody was a vegetarian and lived in awe of the awesomeness of math. So, now you know!

OVERACHIEVER ☼ BADGE

THE WONDERFUL WORLD OF "PYTHAGOREAN TRIPLES"

If you're an overachiever, you're going to love these sections. While you don't have to know all this stuff, you'll definitely be glad that you do!

Clearly, Knot Dude's 3–4–5 triangle is special. But how special is it? I mean, are there actually any other combinations of three whole numbers that can satisfy $a^2+b^2=c^2$? Maybe there just aren't any others. Well, actually, there are others— plenty of them. In fact, there are an infinite number of them! And I can say this without even having to plug in a single number to check. How can I be so sure? Take a look at this:

Cont'd

Cont'd from page 51

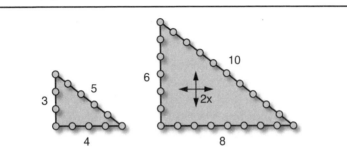

Do you see why now? The only difference between the left and right triangles is that I've scaled the one on the right up to be twice as big as the one on the left. In other words, the legs that were 3 and 4 units long on the left are doubled to be 6 and 8 units long on the right, and so on. And that means that all the sides of the bigger triangle are still whole numbers. So, therefore, Knot Dude's 3–4–5 triangle is *not* the only one of these special right triangles with whole number sides.

But does the bigger triangle on the right still work with the Pythagorean theorem? It had better, but let's try it out just to be sure. For this triangle, $a = 6$, $b = 8$, and $c = 10$. So, $a^2 + b^2 = 6^2 + 8^2 = 36 + 64 = 100$. And $c^2 = 10^2 = 100$. Thank goodness, $a^2 + b^2 = c^2$. It works! If you think about it, you'll see that instead of doubling the sides of the triangle, we could have multiplied them all by 3, 4, or any other whole number too. Each of these scaled-up triangles would have satisfied the Pythagorean theorem, and they all would have had whole number sides.

By the way, all of these special right triangles with whole number sides we've been talking about are known as "Pythagorean triples." So the three numbers (3, 4, 5) describe a Pythagorean triple, as do (6, 8, 10), (9, 12, 15), and so on. But, in truth, these scaled-up Pythagorean triples we've talked about so far are all kind of trivial cases. And by that I mean

Cont'd

Cont'd from page 52

that they're not really very interesting since once you know about one of them, you know about all of them—in other words, "Big whoop!"

So, are there other Pythagorean triples besides (3, 4, 5) that aren't "trivial"? Yep, it turns out that there are. And, again, it turns out that there are an infinite number of them. One way to find them is just to start plugging in numbers. The first triple of numbers you'll find after (3, 4, 5) is (5, 12, 13) since $5^2 + 12^2 = 13^2$. We can check this out to make sure it really does satisfy the Pythagorean theorem: $5^2 + 12^2 = 25 + 144 = 169$, and $13^2 = 169$. So it works! But looking for Pythagorean triples by brute force like this is pretty tedious work. And, as it turns out, there is an easier way. But that's a topic for another day.

SECRET AGENT (in Training) MATH-LIB

WHAT IS PYTHAGORAS' MESSAGE?

Good evening, agent in training _____ *(name of fruit or vegetable)*. Your mission today, if you choose to accept it, is to figure out the _____ *(silly adjective)* secret message that Pythagoras left embedded in his name for new math secret agents to _____ *(unnecessary adverb)* discover. If you succeed, you will graduate from secret agent math academy and become a full-blown secret agent. _____ *(overly enthusiastic exclamation)*!

So, what's in store for your final training session? Well, you might not be surprised to learn that it has to do with one of dear Pythagoras' most _____ *(judgmental adjective)* discoveries: the Pythagorean theorem.

Here's how it's going to work. First, I need you to write down the whole numbers that correspond to the positions in the alphabet of each letter of Pythagoras' name. For example, "A" corresponds to the number 1, "B" to 2, and so on. So, since the first letter of Pythagoras' name is "P," the first number in your list is 16. Write down each number above the corresponding letter here:

$$\underset{(P)}{\overline{16}} \quad \underset{(Y)}{\overline{}} \quad \underset{(T)}{\overline{}} \quad \underset{(H)}{\overline{}} \quad \underset{(A)}{\overline{}} \quad \underset{(G)}{\overline{}} \quad \underset{(O)}{\overline{}} \quad \underset{(R)}{\overline{}} \quad \underset{(A)}{\overline{}} \quad \underset{(S)}{\overline{}}$$

Now, here comes your real secret agent math test. Remember the Pythagorean theorem? (How could you forget!) It's a little algebraic relationship that says: $a^2 + b^2 = c^2$. In the problems below, I give the lengths of the two legs (a and b) and the hypotenuse (c) of six right triangles that satisfy the Pythagorean theorem. Well, almost. I don't actually give you the lengths of all three sides of each triangle—I only tell you the lengths of two of the three sides. Your job is to figure out the length of the third side, and then to match that to the corresponding letter from the name "Pythagoras" above.

For example, in the first problem that follows, the values $a=12$ and $c=20$ are given. Your task is, therefore, to find the value of b. If you do the calculation, or just notice that this must be a scaled-up version of the 3–4–5 triangle (all the sides are multiplied by 4), you'll find that $b=16$. What letter does that correspond to in the name "Pythagoras" above? The very first letter: "P". So you then write that in the space to the right of the problem below. Got it? Good. Remember to look for problems that are just scaled-up versions of the 3–4–5 triangle, and to look for problems that are repeats of ones you've already solved but that ask you to find a different one of the three sides. (Hint: number 4 below looks awfully similar to number 1, doesn't it?) Now have at it!

$$a^2 + b^2 \;=\; c^2$$

1. $12^2 + \mathbf{\underline{16}}^2 = 20^2$ .... **P**

2. $7^2 + 24^2 = \underline{}^2$ .... _____

3. $8^2 + \underline{}^2 = 17^2$ .... $\underline{}$

4. $12^2 + 16^2 = \underline{}^2$ .... $\underline{}$

5. $\underline{}^2 + 15^2 = 17^2$ .... $\underline{}$

6. $\underline{}^2 + 24^2 = 25^2$ .... $\underline{}$

If you've got those six problems solved and six letters from the name "Pythagoras" written on the right, then you have my heartiest congratulations! Now, I've got three more problems for you—but these ones are a lot easier. Just take the numbers that are the solutions to the six problems above, and use them to do the following arithmetic problems. For example, "(5)" and "(6)" indicate that you should substitute in the numerical solutions for problems 5 and 6 above.

7. $(4) - (5) + (6) = \underline{}$ .... $\underline{}$

8. $(2) - (6) = \underline{}$ .... $\underline{}$

9. $(5) - (6) = \underline{1}$ .... $\underline{A}$

Excellent—you've done it! All that's left to do is decipher the secret message with advice for aspiring math secret agents that Pythagoras left hidden in his name thousands of years ago. What's the secret to finding that? Just fill in the letters in the following message that correspond to the solutions of the indicated problem above.

Pythagoras' secret message for math secret agents is…

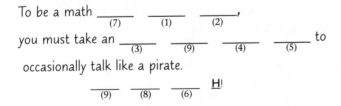

Okay, not quite what you were expecting, right? Well, nobody ever said it was going to be *good* advice—perhaps Pythagoras was getting a little eccentric in his later years. Also, just in case you're wondering what I mean that Pythagoras left this message for you "in his name," the letters in the words "SPY," "OATH," and "ARG(H)" (that last "H" was a bit of a cheat) can be rearranged to spell "Pythagoras." That's right—it's what's known as an anagram!

Congratulations! As of this moment, you have officially graduated your training program and are now a full math secret agent. Good luck in the field, agent _____ (name of fruit or vegetable).

WHY ALGEBRA MATTERS IN THE REAL WORLD

So what's the point of our tale of Knot Dude and Papa Knot? Well, it's certainly not to suggest that this is how things actually went down 4,500 years ago. No, the real point is that something happened in this story that also happens all the time in the real world—and it's a very important something. What is it? It's that some creative person spent time (perhaps too much) thinking about seemingly abstract and crazy things like: How many ways are there to tie knots in a piece of rope? And what interesting arrangements can I make from them? In other words, some creative person spent time thinking about pure abstract math. And, as we saw, every now and then these pure abstract thoughts make their way into the real world.

Typically what happens is that somebody realizes they have a problem that they don't know how to solve. Perhaps, like in our story, they need to figure out how to build a great pyramid that has perfectly square corners. But our story was not typical (in more ways than one) since Papa Knot was lucky enough to have a budding mathematician with an answer to his problem living right under his own roof. Instead, what usually happens is that a physicist, biologist, economist, or some other person trying to understand something seeks the help of a mathematician. They ask the mathematician if they've ever seen anything like their problem before, and the answer is usually one of two things:

- *"No, let me think about it."*
- *"Yeah, actually—I played around with something a while ago that behaves exactly like that. Have a seat, let me show you how it works . . ."*

But this process doesn't just help out mathematicians, engineers, and scientists. Believe it or not, it has even helped you out! Think that sounds crazy? Okay, well it turns out that people didn't invent the mathematical concept of a greater than or less than inequality just to help you figure out if you have enough money to buy the ingredients for a pizza. Nope! What actually happened was that a long time ago people started pondering inequalities and the relationships between numbers simply because they thought it made for a pretty good time. And, lucky for you, all these millennia later, those ideas turned out to be useful!

So there you have it—that's the story of how the abstract world of math that mathematicians spend lots of time in, and which we dabbled with in the prologue, runs smack into the real world. And this, math fans, is also precisely the point at which math moves from the clouds down to the gritty streets and into the realm of "reality." In other words, this is the point at which abstract math is transformed into the math that you actually apply and use in your daily life. Speaking of which, here's a real-world situation for you to think about . . .

M A T H   B R A I N   G A M E

HELP, MY DECK IS CROOKED!

Imagine you've just paid someone to build a new deck in your backyard. The design was simple—it was just supposed to be a 15-by-15-foot square. But, unfortunately, when all was said and done, your simple square deck turned out looking like this:

Top View of
Your Deck

And that's most definitely *not* a square! So what went wrong? Well, clearly the folks building your deck hadn't heard the story of Knot Dude, Papa Knot, and the Great Pyramid of Giza. Otherwise, they would've used a piece of rope with knots in it to help build a deck with square corners. After seeing the deck, you're so angry that you decide to tear it out and rebuild it yourself—properly this time! But while you're fumbling around trying to make a triangle like Knot Dude's, you come to the realization that rope is pretty floppy and isn't a very practical tool. So you stop and think for a while . . . and you come up with something ingenious: a way to use your rope to build a square deck that doesn't require tying a single knot!

HOW TO MAKE A SQUARE —METHOD ONE

So, can you figure out what the "trick" is? If the answer isn't jumping out at you (and it very well might not be), try backing up and rethinking the situation from the beginning. You've got a bunch of rope and four long boards that have been nailed together into a squished square, and your task is to unsquish the square until you have an actual square. Take a few minutes to think about how you could do it (it may help to try sketching out your solution on paper).

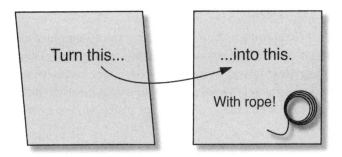

Don't worry if you're not getting it. Believe it or not, you're not really supposed to figure these things out right away. That's right—they're not supposed to be easy! Taking a few minutes to think about the puzzle, and then walking through the solution with me is the key to getting your brain thinking mathematically—and that's our real goal! Eventually the answers will just start to jump out at you.

So, got any ideas? If not, here's a hint: start by thinking about what makes a square a square. So, what is it? Well, first and foremost:

1. The lengths of all four sides have to be equal.

It's absolutely impossible to make a square otherwise! And this means that you'd first better be certain that the lengths of each of the four boards that make up the frame of your deck are exactly the same length. That means you should start by using your rope to measure the length of one side of the deck, and then checking that each of the other three sides match this length.

But will that alone ensure that you've got a nice square deck? Well, take another look at the squished square shape of your crooked deck. It's not a square, but there's no reason that all four sides can't be the same length, right? So the answer must be "no"—four equal sides doesn't guarantee anything. Which brings us to the second critical thing that makes a square a square:

2. All four corners must form right angles—just like
the corners of a pyramid.

Is that it? Do those two things guarantee a square? Yes, that really is it. Go ahead and think about it for a minute and you'll see that if both of these things are true, then you're guaranteed to get a square. Are you now thinking, "Okay, that's great. But how do we actually use that second requirement to make our deck square? I mean, it's nice to know what makes a square a square, but how exactly do I build one?"

Oh right . . . that's a pretty important detail. Well, given all that we've discovered about squares, one option would be to create one of Knot Dude's triangles with a right angle and use it to check that each of the four corners of the deck are square. But you say, "That's exactly what we didn't want to have to do in the first place because Knot Dude's triangle rope is a major pain to work with!" So what can we do?

HOW TO MAKE A SQUARE—METHOD TWO

Okay, it's time to reveal the key idea that will help us move forward and stop talking in circles. While having equal length sides and right angles are all that is required to guarantee that a square is really a square, these properties are not unique. "Whoa!" you might be saying, "What does that mean?" Well, it means that there are other sets of properties that will also guarantee that our square is a square. Yep, sorry to break it to you, but the world is a complex place . . . even the seemingly simple world of squares. Which means that rethinking how you define what makes a square a square might open up a new way for you to fix your deck. In particular, think for a minute about this pair of properties and try to figure out whether or not they guarantee a square:

1. The lengths of all four sides must be equal.
2. The lengths of the two diagonals stretching from opposite corners of the square must also be equal.

What do you think? Do these two things do the trick? The answer is . . . yes!

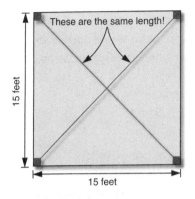

Actually, the fact that the two diagonals have equal lengths can be used to show that each of the four corners also must form right angles, and vice versa . . . just one of the many things you can learn from geometry. Which means that the two sets of properties we came up with are actually different ways of saying the exact same thing. If you're having a little trouble seeing why the two diagonals have to be the same length in a square, take a look at the differences in the diagonals of these two increasingly "squished" squares.

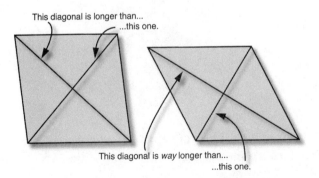

So that's the answer to our puzzle. All you have to do to fix your deck is use your rope to measure the length of one of the diagonals from corner to corner, and then check to see if the length of the other one is the same. If it is, you're done! If not, you need to make an adjustment to further unsquish the square, and then remeasure the diagonals to check your progress. In the end, when all is said and done, the two diagonals of your perfect deck must be exactly the same length. And there you have it—yet another example of the very real way that abstract mathematical ideas can help solve problems in the real world of your own backyard.

ALGEBRA IN YOUR BACKYARD

But wait, there's algebra in the deck puzzle we just solved too. By now you're probably getting the impression that algebra pops up everywhere in life . . . and that's pretty much true! For example, let's take another look at your newly rebuilt and now perfectly square deck.

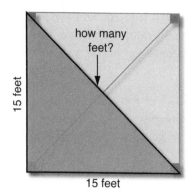

If you just look at half of your deck, what do you see? That's right . . . it's a right triangle! The lengths of the legs of this triangle, *a* and *b*, are both 15 feet, and we can find the length of the diagonal stretching from one corner of your deck to the other using the good old Pythagorean theorem. How? Well, first we know that

$a^2 + b^2 = 15^2 + 15^2 = 225 + 225 = 450.$

And since $a^2 + b^2 = c^2$, this means that

$c^2 = 450.$

Finally, we need to take the square root of 450 to get the value of c. Which, after running the numbers through a calculator, gives us a number that's approximately equal to 21.21.

So what's the point of this whole discussion? Well, the point is that we can use a little bit of algebra to learn that a perfectly square deck with two equally sized 15-foot-long sides *must* have a corner-to-corner length of about 21.21 feet. And that means that instead of using a rope to check that the two diagonals of your deck are the same length, we could instead

- Measure the width and length of the deck using a tape measure.
- Then use these lengths in the Pythagorean theorem to calculate the length of the hypotenuse stretching from corner-to-corner of the deck.
- Finally, use the tape measure to check and make sure that the length of the deck's hypotenuse is correct.

Either method will work and ensure that your deck ends up square. That's some pretty handy math, right?

• ALGEBRA TUTORIAL •

HOW TO MAKE A GRAPH

Throughout the book, you'll find sections like this containing "algebra tutorials." Each of these sections gives an in-depth look at a particular type of algebra problem and includes a step-by-step guide showing you how to solve them. Some of the steps are left for you to do, but if you get stuck you can find solutions at the end of the book in Math Dude's Solutions.

While using the Pythagorean theorem to find the length of the diagonal line stretching from corner-to-corner of your 15-by-15 foot square deck is nice, we can do a lot more with what we've learned so far than that. For example, imagine that you build decks for a living. A lot of the decks you build are square, but most of them are not exactly 15-by-15 feet in size like the one we just looked at. Some of them are 21-by-21 feet, some of them are 12-by-12 feet, and so on. As a deck builder, it'd be useful if you had some way of quickly looking up the size of a deck and then immediately seeing what its diagonal length should be . . . no matter how big or small it is. How can we do that? Well, the first thing we need to do is learn how to make a graph.

■ STEP 1: DRAW CARTESIAN COORDINATES

Let's start by talking about how to get set up so that we're ready to make a graph. Which means that we need to draw ourselves a pair of coordinate axes, like these:

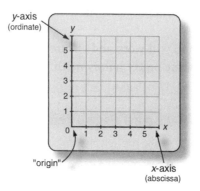

What exactly are we looking at? Well, this is what's called a Cartesian coordinate system (named after the seventeenth-century mathematician René Descartes), and it's sitting on top of what's called the coordinate plane. Its main purpose is to give us a way to investigate the relationship between numbers (more on exactly what this means later). In this case, we want to look at the relationship between pairs of numbers, which means that we need two axes in our coordinate system. And that's exactly what we have: the x-axis (sometimes called the abscissa) runs horizontally and the y-axis (sometimes called the ordinate) runs vertically. The little marks along each axis label the x and y values at those locations. Both axes have a value of 0 at the origin—the place where the two axes meet. So that's our playing field . . . now let's start doing something on it.

■ STEP 2: CREATE LIST OF ORDERED PAIRS

In case you haven't guessed it by now, our plan for coming up with a way to quickly figure out the diagonal length of any sized deck is centered around the idea of making a graph. But before we make that graph, we need to do a little algebra with the Pythagorean theorem. Our goal in this problem is to figure out the length of the hypotenuse, c, of the triangle formed by half a deck:

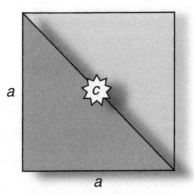

Since we want to solve for c, let's start by switching around the left and right sides of the Pythagorean theorem and write it like $c^2 = a^2 + b^2$. Then, let's take the square root of both sides to get

$$c = \sqrt{a^2 + b^2}$$

But notice that there isn't a side called "b" in our drawing of the deck. Instead, both sides are labeled "a" since this deck is shaped like a square—which means that both sides have the same length. So let's use the fact that $a = b$ for this deck to rewrite our equation like this

$$c = \sqrt{a^2 + a^2} = \sqrt{2a^2} = \sqrt{2} \cdot \sqrt{a^2} = \sqrt{2} \cdot a$$

Don't worry if you don't understand every single step here yet—we'll go over solving this type of problem in much more detail in the next chapter. But for now the really important thing to notice is that the length of the hypotenuse of our deck is given by $c = \sqrt{2} \cdot a$

Which means that we can use this equation to figure out what the diagonal length of *any* square deck will be . . . we simply plug in values for a, and in return we get values for c. For example, if we do this for $a = 1$, we get $c = \sqrt{2}$ If we do it for $a = 2$, we get $c = \sqrt{2} \cdot 2$—which is more often written $2\sqrt{2}$ (we can omit the "•" since it's clear we're talking about multiplying 2 by $\sqrt{2}$). And if we do this for a bunch of values of a, we end up creating a list like this one:

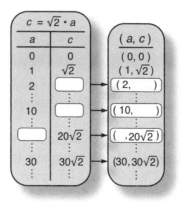

$c = \sqrt{2} \cdot a$		(a, c)
a	c	
0	0	(0, 0)
1	$\sqrt{2}$	(1, $\sqrt{2}$)
2	☐	(2,)
⋮	⋮	⋮
10	☐	(10,)
⋮	⋮	⋮
☐	$20\sqrt{2}$	(, $20\sqrt{2}$)
⋮	⋮	⋮
30	$30\sqrt{2}$	(30, $30\sqrt{2}$)
⋮	⋮	⋮

I'll leave it up to you to fill in all the missing parts. But what is all that stuff on the right? That sure is an odd way to write a list of numbers! Actually, those are called ordered pairs . . . and they're exactly what we need to make our plot.

■ STEP 3: PLOT POINTS ON THE PLANE

Now we're ready to plot some points. But why would we want to do that? Well, as I mentioned earlier, plots help us see the relationships between numbers. Each point on what's called the *x-y* plane (which is just another name for the Cartesian coordinate system we set up) is defined by a single pair of numbers. So by plotting several points, we can begin to investigate how these pairs of numbers are related to each other. How does this all work? Well, before we tackle our deck problem and the ordered pairs we just came up with, let's take a look at this graph:

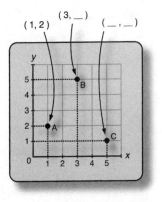

The positions of each of the three points plotted on this graph are described by an ordered pair of numbers. In exactly the same way that a pair of cross streets directs you to a location on a map, an ordered pair of numbers tells you the location of a point on the coordinate plane. All you have to do to describe a location on the plane is to give a location along both the x and y axes. Ordered pairs are typically written inside parentheses with the x coordinate given first. For example, the location of the origin is specified by the ordered pair (0, 0), the location of the point labeled "A" at $x=1$ and $y=2$ is specified by the ordered pair (1, 2), and the location of the point labeled "B" at $x=3$ and $y=5$ is . . . well . . . I'll leave it to you to finish up writing this ordered pair. And while you're at it, go ahead and fill in the ordered pair for point C too.

Okay, after doing that you should now understand exactly what those ordered pairs that we came up with earlier were all about! They simply describe the locations of points on a graph. But which graph? Well, this one:

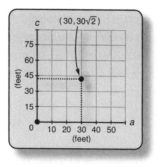

What you see here is the start of what will eventually become the graph that helps us solve our deck problem. Notice that instead of an x-axis we have an "a-axis," and instead of a y-axis, we have a "c-axis." Why? Because we don't have (x, y) ordered pairs, we have (a, c) ordered pairs. In other words, since the table we made earlier contains a and c values, the ordered pairs we get from it are (a, c) ordered pairs. So far I've plotted the ordered pairs (0, 0) and (30, $30\sqrt{2}$). I'll leave it up to you to plot the rest of the points from the table we made earlier. And while you're at it, go ahead and add the two points for $a=40$ and $a=50$ that aren't in the table. Feel free to use a calculator to help you figure out the approximate values of all those numbers multiplied by $\sqrt{2}$. Once you're done with all that, you'll be ready to move from plotting points to plotting curves.

■ STEP 4: CONNECT THE DOTS TO DRAW A CURVE

All finished with slapping those (*a, c*) ordered pairs on the plot? Okay, there should be at least six points on your graph showing the locations of the ordered pairs corresponding to deck sizes of *a* = 0, 10, 20, 30, 40, and 50 feet (you might have some additional points too). While plotting these points, you should start to see a pretty clear trend developing that tells you about how these points are related to one another. In this case, you should be able to see that the points appear to be ascending in a straight line from the lower left to the upper right of the plot. So, let's go ahead and draw a line that connects the dots and shows the overall trend. We'll learn more about this in chapter 3, which will help you understand why the points in this problem end up in a straight line.

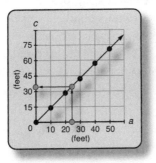

Okay, so what does this graph do for us? Remember way back in the beginning of this tutorial when we said that we wanted to come up with a way to quickly find the diagonal length of a square deck given only the length of its sides? Well, that's precisely what this graph does for us! For example, let's say you're building a 24-by-24-foot deck. You can easily find the diagonal length of this deck by starting at *a* = 24 feet on the "*a*-axis," then moving vertically up from this point until you hit the diagonal line that we drew, and finally moving horizontally to the left until you hit the "*c*-axis." The value of *c* at that point is the diagonal size of the deck. With just a quick glance at the graph we can see that this value must be a bit more than 30 feet (the actual number is $24\sqrt{2}$ ... which is close to 34 feet). So, in the future, when you need to make a graph, just remember to follow the four steps we used in this tutorial:

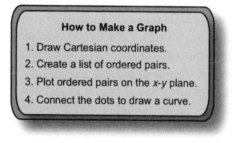

How to Make a Graph
1. Draw Cartesian coordinates.
2. Create a list of ordered pairs.
3. Plot ordered pairs on the *x-y* plane.
4. Connect the dots to draw a curve.

Before finishing, check to see if you can figure out the approximate diagonal lengths of square decks with 15- and 45-foot-long sides (bonus points if you can also calculate a more precise estimate using a calculator). Once you've got that all figured out, you're done with this tutorial! But don't worry, while this graphing tutorial may be over, rest assured that we're nowhere near done with making plots . . . not by a long shot.

POP QUIZ: CONNECT THE DOTS

Speaking of doing more graphing, I think it's high time for you to make a graph of your very own . . . from scratch. Here's what I want you to do: Make a graph showing the relationship between the perfect squares and the whole numbers you multiply together to make them. In other words, make a graph showing the ordered pairs of points (n, n^2). Don't worry, n here isn't anything crazy—it's just representing all

Cont'd

Cont'd from page 70

the various whole numbers. So the first ordered pair for $n = 1$ is (1, 1), the second ordered pair for $n = 2$ is (2, 4), and so on. After you come up with a list of ordered pairs (going up to $n = 10$ should do the trick), plot the points on the axes, and then draw a curve that goes through the points and shows the overall trend.

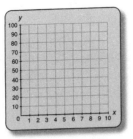

Bonus question: What does this curve you've drawn mean? In other words, all of the perfect squares are sitting nicely on the points you plotted, so what do the parts of the curve that are located between the points represent?

CHALLENGE PROBLEMS

Okay, now that we have a better understanding of the big picture of what math really is and how it gets applied to the real world, and now that we've put together an impressive collection of mathematical tools for our journey, it's time for us to head out and start our trek into the real bread and butter of this book. And trust me, that's definitely a journey you want to take. Because when we're finished you'll be able to do things that you simply can't do right now. For example, here are three problems that, barring some painful brute-force expenditure of effort, you simply can't solve right now:

1. Pick a whole number greater than 2. Now multiply it by 2, and then subtract it from the square of your original

number. Then add 1 to this result. After that, take the square root of this number, and then add 1 to the result. What's the answer? After all that work, did you get back exactly the same number you started with? How can that be? Magic, right? No, it's algebra.

2. You are told that the product of two consecutive whole numbers is 1056. Can you figure out what those two numbers are? If you think about it, this seems like a nearly impossible problem to solve without simply trying to multiply a bunch of numbers together until you get the right pair. But how can we do it smarter and faster?

3. You drop a rock into a well and hear a splash two seconds later. Can you figure out how far it is from the top of the well to the surface of the water? There's a lot going on here. We have to worry about the physics of gravity, the motion of the rock, the speed of sound, and how this all comes together mathematically. Do you think we'll be able to do it?

By the end of this book, I assure you that we will be able to solve all of these problems . . . and many more too. So, I'd like to formally challenge you to keep going until we solve them. Of course, before we can do that, there are a few things we need to learn about: variables, expressions, equations, exponents, polynomials, factoring, quadratics, and more. In other words, we need to learn algebra.

So, have a seat, let me show you how it works.

· FINAL EXAM ·

• ALGEBRA IN THE EVERYDAY WORLD

1. We've seen how algebra can show up in our shopping carts, our monthly budgets, and even in our backyards. What other ways does algebra enter into your daily life? Take a few minutes and

think about it. It doesn't matter how big or small your list is—what really matters is that you start to recognize the math that's lurking in your life . . . and then embrace it!

- **INEQUALITIES**

2. Fill in the blanks with >, <, ≥, ≤, or = to make each statement true.

- three perfect squares past 1 ____ 8 / (6+2) • 17
- (2+8) • 10 ____ (2+8) / 10
- 2+5 ____ 5+2
- (8+2) / 10 ____ (8+2) • 10
- (9 • 2) • 8 / (6+2) ____ three perfect squares before 49

Some of these questions can have more than one correct answer. Can you figure out which those are?

- **SQUARE ROOTS**

3. What's the square root of a square? Um . . . huh? Okay, I know this one sounds a little strange, but here's what I'm talking about:

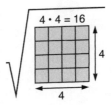

$4 \cdot 4 = 16$

4

4

In other words, the big square here is 4 little boxes wide and 4 little boxes tall, which means that it contains a total of 4 • 4 = 16 little boxes. So, what's the square root of a square?

- **THE PYTHAGOREAN THEOREM**

Building square decks isn't the only backyard problem that algebra—and in particular the Pythagorean theorem—can help you with.

Imagine that you need to buy a ladder to climb onto the roof of your nine-foot-high house. How tall of a ladder should you buy? Should it be nine feet tall?

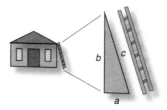

After seeing this picture, it's clear that a nine-foot-tall ladder won't do since the ladder is going to form the hypotenuse of a triangle. So, high tall should it be? Well, the answer will depend on the height of the roof, *b*, and how far you need to position the bottom of the ladder from the house to keep it from falling over, *a*. Given all that, can you figure out the answers to the following two questions?

4. If the height from the ground to the roof is nine feet and you decide that the bottom of the ladder should be positioned three feet from the house, how tall of a ladder do you need to buy to be able to get on the roof?

5. You have a twelve-foot-tall ladder in your garage, and you've determined that the bottom of the ladder should be positioned a distance one-third the height of the ladder away from the house. How tall of a roof can you climb on using this ladder?

• GRAPHING POINTS AND LINES

6. Plot the two relationships $y_1 = x$ and $y_2 = 2 \cdot x$ on the same set of axes. In other words, first choose a bunch of *x* values (the

whole numbers from 1 through 10 will do), plug these x values into these two relationships to find ordered pairs for (x, y_1) and (x, y_2), plot these points (use different symbols for each relationship), and then draw lines connecting the dots. What effect does the 2 in y_2 have on the result?

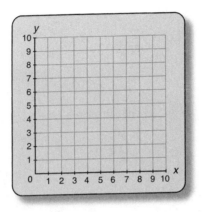

CHAPTER 2

Algebra Basics

> He say one and one and one is three...
>
> —The Beatles, "Come Together"

People often think that "algebra" sounds like a big and scary thing, but that's just because they don't understand what algebra really is. So, the first thing we're going to do is tackle this problem head-on and answer what just might be the biggest question of all:

What the heck is algebra?

The problem is that algebra isn't a simple object . . . like a spoon. If it was, it'd be easy to explain. After all, few people who eat lunch with a spoon will have trouble explaining how it works afterwards. But, unfortunately, that's just not what algebra is. So, what is it then? Well, instead of being a thing like a spoon, it's more like a style or a method of using a spoon . . . I'm thinking of a sensible two-fingered (plus a thumb) grip versus a crazy five-fingered silverware stranglehold. Sure, both styles will get the job done, but one technique is definitely more efficient (and sophisticated) than the other.

The same is true with "style" in math. One type of method—

sometimes it's algebraic—is usually most appropriate for solving certain types of problems. But styles are much harder to recognize, point at, and label than things—which is why recognizing and labeling algebra can be tricky. So, what's the best way to think of algebra then? My advice is to think of algebra like this:

Algebra is arithmetic . . . with variables.

Honestly, once you understand that, you understand algebra. And the good news is that you're probably pretty proficient at doing arithmetic already. So you just need to focus in on becoming comfortable with the one key caveat in that description . . . the _"with variables"_ part.

WHAT ARE VARIABLES?

Okay, so if algebra is arithmetic with variables, we'd best take some time getting to know about variables. We've seen several variables in action already. For example, _n_ from our foray into the world of perfect squares and _a, b,_ and _c_ from our adventures with the Pythagorean theorem were all variables. But, in general, what are variables?

Variables are symbols without predetermined values.

I know this sentence might look like a massive mess of weirdness right now, so let's sort it all out and try to make some sense of it.

SYMBOLS IN MATH

First, let's talk about the many different types of symbols used in math. Here are a few that you're already quite familiar with:

- "+" and "–" symbolize ways to combine numbers (there are others too)
- ">" and "<" symbolize ways to compare numbers (again, there are also others)

- 1, 2, and 3 symbolize the concept of a number (and, of course, there are *many* others)

You might be surprised by this last type of mathematical symbol. Most people don't differentiate between the concepts of the numbers 1, 2, 3, and so on, and the symbols (called "numerals") used to describe them. But the ideas of 1 and 2 definitely have lives and meanings far beyond the way we write "1" and "2." For example, the number 2 has an abstract property of "two-ness" associated with it. It doesn't necessarily mean two people, two seconds, or two anythings—it just means 2. That is, it just means two-ness.

NUMERALS

Of course, we could write our symbol to represent the idea of two-ness in many different ways. Most folks are now fans of using what's called the Hindu-Arabic numeral system (for reasons we'll discuss in the next chapter) and express the idea of two-ness using the numeral "2." But we could just as easily have used a different symbol—perhaps two dots like "• •" or two vertical lines like " | | ." In fact, the latter is exactly what the Romans used to represent 2 in their numeral system a few thousand years ago.

These boxes contain the **idea** of "two-ness" that we write "2."

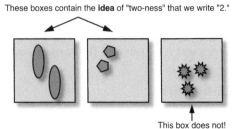

This box does not!

The important thing to notice is that a symbol like "5" has a predetermined and universally agreed upon numerical value. In other words, it always means ". . . . ." dots.

The fact that we've associated numerical values with symbols gives us the power to do a bunch of handy things . . . like add and subtract numbers. For example, you can think of the problem $3 + 5 = 8$ as being a bunch of symbols that describe the "real" problem, which is trying to figure out how many total dots you get when you put

••

into the same box as

•••••

In other words, $3 + 5 = 8$ is a collection of symbols that represents the real problem:

••• + ••••• = ••••••••

Clearly, both ways of stating the problem give the same solution, but it's also clear that the version that uses symbols and numerals (that's the one written $3 + 5 = 8$) instead of dots is much easier to work with. We'll talk more about this in the next chapter.

POP QUIZ: THE ADVANTAGE OF USING DOTS?

Can you think of any advantages that dots have over numerals when it comes to doing arithmetic? **Hint:** You might find it useful to think about the solution we came up with to the "positively odd" Brain Game from the prologue.

Numerals like 1, 2, and 3 are a special type of what's called a "mathematical constant." In other words, they're symbols that represent values that don't change: "1" always represents "•," "2" always represents "• •," and so on. People don't often think of numerals as being constants, but they really are. What people normally mean when they talk about a mathematical constant is a number like

$$\pi = 3.14 \ldots$$

This number is called "pi" (π is a letter of the Greek alphabet), and it's probably the most famous constant of them all.

OVERACHIEVER ✨ BADGE

WHAT IS π? AS IN PI, NOT PIE.

Where does the mathematical constant π come from? To see, you first need to draw a circle. Now draw a line going from one side of the circle to the other through the circle's center. The length of this line is called the circle's "diameter." Next, "cut" the circle at the top, stretch it out, and lay it flat. The length of this line (which is the total length around the circle) is called the circle's "circumference." Now take the diameter line that you drew earlier and count how many times it fits along the outstretched line of the circle's circumference.

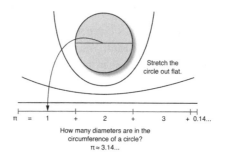

Stretch the circle out flat.

π = 1 + 2 + 3 + 0.14...

How many diameters are in the circumference of a circle?
$\pi \approx 3.14...$

What's the magic number? Well, it turns out that a little bit more than 3 diameters of a circle fit along its circumfer-

---

Cont'd

Cont'd from page 80

ence. A better estimate of the number of diameters is 3.1, but that's still not quite enough. An even better estimate is 3.14, but again that's still not quite enough. And you can keep on getting better and better approximations by adding more and more decimal digits: 3.141 . . . , 3.1415 . . . , 3.14159 . . . , and so on. But π doesn't have an exact decimal value—the digits keep on going forever without forming a repeating pattern. Numbers like this are called "irrational," and we'll talk more about them in the next chapter.

FROM SYMBOLS TO VARIABLES

Okay, let's summarize where things stand at this point. So far we've talked about three things:

- Algebra is arithmetic . . . with variables.
- Variables are symbols *without* predetermined values.
- Numerals are symbols *with* predetermined values.

So what does this all mean? Well, let's work backward through the list and unravel the meaning of each thing one at a time. First, starting at the end, numerals are just symbols with set in stone values (perhaps literally). In other words, the idea of "two-ness" or "three-ness" is encapsulated by the symbols "2" and "3." Now, with that in mind, it's not such a big leap to understand what a variable is. Just like numerals, variables are symbols that represent numbers . . . except that the numbers they represent aren't yet specified. Honestly, that's the only difference between something like 5 and something like x!

SO WHAT, EXACTLY, IS ALGEBRA?

We'll come back to this in a minute, but let's assume that it all makes sense. Then, working up to the first item on the list, we see that we now know and understand all of the pieces that are required to "get" algebra. Here's the gist:

Algebra is exactly like good old arithmetic . . . except that instead of just doing arithmetic with numbers, we're also going to do it with these new things called variables, which are just symbols that represent the as yet unknown values of some numbers.

I know it's a mouthful, but now that you've seen it put this way, it's pretty clear that algebra isn't so scary after all, right? Of course, there's a lot more to understand about these things, but that's exactly what the rest of the book is for.

HOW DO VARIABLES WORK?

Okay, let's now take a closer look at how variables work . . . or, rather, how to work with variables. The first thing we should do is pick a symbol for the variable we're going to use in this chapter. As you may know, the letter *x* is used all the time in algebra and, of course, it'd work just fine for us here too. But we're going to be seeing a lot of dear letter *x* later in the book, so let's be a bit more adventurous in this chapter. Instead of *x*, let's choose $\square$.

HOW TO NAME VARIABLES

But why have we chosen to use $\square$ as our variable? Is it special in some way? Well, I'm sure its parents would disagree, but the truth is that $\square$ is not particularly special. In fact, we could have used *any* symbol we could think of, because the symbol doesn't actually matter!

ALGEBRA DECODER!

- *X*

Why is the letter *x* used so often in algebra? In truth, nobody knows for sure. One explanation is that *x* is simply easy to write . . . which was a really big deal back in the day when people were scratching out the written word with feathers and ink. But there's another interesting possibility. As we'll soon discover, the main focus of algebra is to solve for the unknown value of an object (which we call a variable). Well, it turns out that the Arabic word for "object" (the original algebra book was written in Arabic) is "shei," which when translated into Greek is "xei." All it takes then is one little abbreviation of "xei" down to "*x*" and, voilà, you've got an explanation. Well, maybe.

POP QUIZ: WHAT'S THE BEST VARIABLE NAME?

As we've discussed, you're free to use whatever names you want for your variables. But that doesn't mean you should actually use absolutely any name at any time. No, some choices are definitely better than others for certain quantities. For example, it's common in physics to use the variable names E for energy and m for mass—as in $E = m\,c^2$ (you may have heard of that one). The E and m are convenient here because they serve as a quick reminder of what you're looking at. With that in mind, can you come up with some good variable names for the following quantities?

1. The length of your new sailboat (for calculating if it will fit in your garage)
   . . . . _____

2. The height of the ceiling in your garage (again, to see if the sailboat will fit)
   . . . . _____

3. The weight of your boat (to see if your garage floor will hold it)
   . . . . _____

Cont'd

Cont'd from page 83

---

4. The speed that your sailboat is moving through the water
   .... _____

5. The speed of the wind that's causing your sailboat to move at that speed
   .... _____

   Of course, there's a bit of tradition in variable names too. For example, the variable names x, y, and z, as well as a, b, and c are commonly used as generic names . . . and they're just fine for you to use too.

WHAT CAN WE DO WITH VARIABLES?

Now that we have our variable, $\square$, let's find out what sort of things we can do with it. The short answer is that we can do anything with it that we can do with numbers. For example, one thing we can do is add our variable to the number 3, like this

$$3 + \square$$

and then, afterwards, we can subtract our variable from this sum

$$3 + \square - \square$$

Of course, when we first add and then subtract our variable, the net result is doing nothing—just as it would be if we added and then subtracted 2 from 3. In other words, $3 + 2 - 2 = 3$, and similarly

$$3 + \square - \square = 3$$

So the answer is just 3 . . . no matter what the value of $\square$ is. Okay, what else can we do? Well, among other things, if we want to we can add our variable to itself, like this

$$\square + \square$$

or multiply it by a number, like this

$9 \cdot \square$

and even multiply our variable by itself, like this

$\square \cdot \square$

Yes, these may look strange, but I assure you they're completely legitimate. And I can actually prove that to you. How?

HOW TO WORK WITH VARIABLES

We don't have to leave the problems we've written in terms of $\square$. In fact, we're free to assign an actual number to our variable whenever we please. So, just for fun, let's set the value of our variable equal to the number 2. To make this as visual as possible so that you can really picture what's happening, let's imagine that $\square$ is no longer an empty box, but that it's now a box that contains the number 2. When we now add a number to our "variable" (which isn't actually so variable anymore), we get

$3 + \boxed{2} = 5$

And that makes perfect sense, right? It had better make sense because it now looks exactly like normal arithmetic. The only difference is the outline of a box that reminds us that the number 2 is the value of the variable. Okay, how about when we add our variable to itself. Well, in that case we get

$\boxed{2} + \boxed{2} = 4$

Again, this is exactly what we'd expect from basic arithmetic. How about multiplying our variable by a number

$3 \cdot \boxed{2} = 6$

or multiplying it by itself

$$\boxed{2} \cdot \boxed{2} = \boxed{2}^2 = 4$$

Yes, these give perfectly reasonable results as well. The bottom line is that variables are exactly like ordinary numbers. You can add them, subtract them, multiply them, divide them, and do anything else with them that you can do with numbers in ordinary arithmetic. The only difference is that variables don't have numerical values. Admittedly, this fact makes them kind of weird, but it also makes them extremely powerful.

WHAT ARE ALGEBRAIC EXPRESSIONS?

Why do I say that variables are so powerful? Well, it's not really about what they are, it's more about what we can do with them. In particular I'm talking about the fact that we can use them to create algebraic expressions. So what exactly is an expression? An expression in algebra is like a phrase in English. It's a statement that's made of numbers, variables, and operators like "+," "−," "•," and "/." Notice that I didn't say that an expression in algebra is like a *sentence* in English, I said it was like a *phrase*. The honor of being an algebraic sentence is reserved for something else . . . which we'll talk about soon (spoiler alert: sentence= equation).

HOW TO SPOT AN EXPRESSION?

So what exactly does an expression look like? Well, here's an example of an extremely simple expression:

1

Is that it? Yes, that's it. The number 1, the number 2, or anything else like this is technically an expression. Why? Because, like all expressions, it's made of numbers, variables, and operators. Though these expressions are nothing but numbers, not all expressions are so simple. Here's a slightly more complex expression consisting of two numbers and an operator:

$5 + 10$

It's still pretty simple, but we can make things that are even more complicated. For example, here's an expression containing two numbers, two operators, and a variable

$101 \cdot \Box + 1001$

If we were so inclined we could go on and on like this creating ever more complicated expressions. But I think that's enough to get the idea.

POP QUIZ:
ENGLISH TO EXPRESSION TRANSLATION

A lot of people have trouble translating ideas spoken in English into expressions written in algebra. Most often this problem rears its ugly head when people get stuck trying to translate "word problems" into mathematical expressions. We'll look at some problems that fall into this category later on in this chapter, but before we get to those let's take a few minutes to practice translating some simple expressions. For each of the following, your job is to translate the English phrase into an algebraic expression.

1. "three times $\Box$ squared plus $\Box$ minus ten" → _____
2. "square root of five times $\Box$ divided by $\Box$" → _____
3. "two plus eight times six divided by three" → _____

Cont'd

Cont'd from page 87

Did you run into any problems with those translations? Well, you should have! Normally I won't give you "trick" questions without warning you first, but in this case it's instructive to discover the problem for yourself. In case you haven't figured it out yet, the problem is that all of those English phrases can be interpreted in multiple ways. For example, the first problem could be translated as

$$3 \cdot \square^2 + \square - 10$$

but it could also be translated as

$$(3 \cdot \square)^2 + \square - 10$$

Do you see why either of these are correct translations? This problem makes it pretty clear that algebra is much better suited to talking about math than English.

But not to worry because there's a perfectly good fix for this problem. Actually, there are two. The first fix is simply to remember that whenever you're communicating algebraic ideas using English, you need to be careful . . . really careful. I was intentionally sloppy with my language in this quiz, and you can see the consequences. But you can avoid the majority of these problems simply by clearly communicating what you're saying and by anticipating any ambiguities. The second fix is called the order of operations, which sounds like a great topic for an algebra tutorial.

• ALGEBRA TUTORIAL •

THE ORDER OF OPERATIONS

What does "order of operations" refer to? Well, it has to do with figuring out which part of an arithmetic problem to solve first. Here's an example: What's the answer to

$2 + 3 \cdot 4 = \underline{\quad}$

Okay, it looks like there are two ways to do it. The first is to do the addition first, like this

$(2+3) \cdot 4 = 5 \cdot 4 = 20$

and the second is to do the multiplication first, like this

$2 + (3 \cdot 4) = 2 + 12 = 14$

But those two answers are different! Which means that one of them must be right and the other must be wrong, right? That's correct. So how do you know which is which? That's where the order of operations comes into play. Here's how it works—when solving an arithmetic problem, follow these steps:

■ STEP 1: PARENTHESES

The **first** step is to do any parts that are inside parentheses. For example, in the problem $(2+3) \cdot 4$, the first thing you do is solve $2+3$. So, $(2+3) \cdot 4$ simplifies to $5 \cdot 4$. Parentheses can look like "( )", "[ ]", or even "{ }". The second style, "[ ]", are often called "brackets," and the last style, "{ }", are often called "braces." Sometimes you'll see more than one style used together when several sets of parentheses are needed in the same problem. For example, you can make a problem like $2 / ((4+3) \cdot 6))$ a little easier to read by instead writing it as $2 / [(4+3) \cdot 6)]$. **In a situation like this with nested parentheses, the plan of action is to work from the inside out.**

■ STEP 2: EXPONENTS

The **second** step is to raise numbers to a power—a process that's known more formally as "exponentiation." One example of exponentiation that we've seen a lot of so far is squaring a number. For example, in the problem $2^2 + 3$, the first thing you need to do is square the number 2. That means that $2^2 + 3$ simplifies to $4 + 3$. Or, in the problem $(3 \bullet 4)^2 + 3$, you'd first do the part in parentheses, as we just talked about, so the problem would then become $12^2 + 3 = 144 + 3 = 147$.

■ STEP 3: MULTIPLICATION/DIVISION

The **third** step is to do any multiplication or division, **from left to right**. For example, in the problem $2 + 3 \bullet 4$, the first thing you need to do is solve the multiplication problem $3 \bullet 4$. That means that $2 + 3 \bullet 4$ simplifies to $2 + 12$. Notice how similar this problem is to the problem we looked at earlier: $(2 + 3) \bullet 4$. However, the answers are different because in that problem we had to do the addition before the multiplication since the parentheses took precedence over the multiplication.

■ STEP 4: ADDITION/SUBTRACTION

The **fourth** and final step is to do any remaining addition or subtraction, *from left to right*. For example, in the problem we just looked at, $2 + 3 \bullet 4$ simplifies first to $2 + 12$, leaving a final addition problem. The answer is, of course, $2 + 12 = 14$.

PEMDAS

Okay, those are all the rules you need to know. You may remember learning these rules using one of a number of mnemonics—the most famous of which (at least if you're from the United States) is: "Please Excuse My Dear Aunt Sally," or "PEMDAS" for short.

P is for Parentheses
E is for Exponents
M and **D** are for Multiplication and Division
A and **S** are for Addition and Subtraction

One thing to notice about "My Dear Aunt Sally" is that **M** and **D** and **A** and **S** are together, since they have the same precedence. In other words, you don't do all multiplication followed by all division—you work from left to right doing either multiplication or division as you encounter it. The same is true of addition and subtraction. These mnemonic devices are helpful, but once you get used to doing arithmetic, this will all just become second nature.

SAMPLE PROBLEMS

Okay, let's put all of this together and look at a few example problems and the procedure you should think through when solving them.

$$32 / (3+1)^2 + 9 \bullet (2-1) = \ldots$$

Step 1—Parentheses .... $32 / (\mathbf{3+1})^2 + 9 \bullet (\mathbf{2-1})$
       (Once we're inside the parentheses, we have to
         start over.)
       Step 1a—**Parentheses** ....
       Step 1b—**Exponents** ....
       Step 1c—**Multiply and Divide** ....
       Step 1d—**Add and Subtract** .... $32 / 4^2 + 9 \bullet 1$
       (We've finished everything inside the parentheses,
         so let's move on!)
Step 2—**Exponents** .... $32 / 16 + 9 \bullet 1$
Step 3—**Multiply and Divide** .... $2 + 9$
Step 4—**Add and Subtract** 11

$10 \cdot (2^2)^2 + 10 \cdot (1+1) = \ldots$

Step 1—Parentheses .... $10 \cdot (\mathbf{2^2})^2 + 10 \cdot (\mathbf{1+1})$
      Step 1a—Parentheses ....
      Step 1b—Exponents .... $10 \cdot 4^2 + 10 \cdot (\mathbf{1+1})$
      Step 1c—Multiply and Divide ..   ...
      Step 1d—Add and Subtract ..... $10 \cdot 4^2 + 10 \cdot 2$
Step 2—Exponents .... $10 \cdot 16 + 10 \cdot 2$
Step 3—Multiply and Divide .... $160 + 20$
Step 4—Add and Subtract 180

PRACTICE PROBLEM

Okay, now that we've got a few examples under our belt, it's time for you to take a shot at solving a problem:

$5 + 4 \cdot 3 / 2 - 1^2 = \ldots$

Step 1—Parentheses .... _____
Step 2—Exponents .... _____
Step 3—Multiply and Divide .... _____
Step 4—Add and Subtract .... _____

That wasn't so bad, right? Finally, the single best piece of advice that I can give you to help keep the order of operations straight is this:

Q&DT ▸

Fear not parentheses.

In other words, use parentheses (and brackets and braces) liberally. They're absolutely free to us, and there's no penalty for using too many of them. Why is it a good idea? Because parentheses show you which steps to perform first and save you from having to figure out the order of operations!

Oh, and just in case you're wondering why "PEMDAS" is "PEMDAS" and not, say, something like "PEASMD" instead (with the precedence of multiplication and division swapped with that of addition and subtraction), the answer is that it's really just because a long time ago some people defined arithmetic to work that way. In other words, the order of operations is just a convention. Like all choices of notation in math, there's no reason it *has* to be that way . . . it just is. And it works, so we all merrily go along with it.

WATCH OUT!

Negative Numbers and PEMDAS

Is -2^2 the same as $(-2)^2$? Or is it the same as $-(2^2)$? If you said the first one, you're right! But why? Doesn't the order of operations say we should do exponentiation before multiplication? Yes, that's true . . . E comes before M. And if the problem had been $-1 \cdot 2^2$, then indeed the correct answer would have been $-(2^2)$. But the problem is instead asking you to square the number -2. In other words, it's asking you to calculate $(-2)^2$.

POP QUIZ: FOUR FOURS

Okay, it's time for another math game. This one is called Four Fours, and the goal of the game is to come up with nine expressions that have values ranging from 1 to 9. But here's the trick, each of the nine expressions must use exactly four 4s and three of the big four arithmetic operators ($+$, $-$, $\cdot$, or $/$)—and nothing else! For example, here's

Cont'd

Cont'd from page 93

one way to write an expression that's equal to 1 using four 4s and three operators:

4 _/_ 4 _+_ 4 _−_ 4 = 1

As you can check, this expression says that first dividing two 4s and then adding and subtracting the next two 4s equals 1. In this challenge, I'll also let you throw in sets of parentheses wherever you'd like so that you can change up the order of operations. For example, here's one way to write an expression that equals 3:

(4 _+_ 4 _+_ 4) _/_ 4 = 3

Notice that the parentheses are necessary here . . . otherwise you'd get a different answer (although that other answer might come in handy for one of the other numbers).

Okay, that's it for the instructions. All that's left for you to do is figure out how to create expressions following these rules that equal the numbers 2, 4, 5, 6, 7, 8, and 9 (since I've already done 1 and 3 for you).

4 _/_ 4 _+_ 4 _−_ 4 = 1
4 __ 4 __ 4 __ 4 = 2
(4 _+_ 4 _+_ 4) _/_ 4 = 3
4 __ 4 __ 4 __ 4 = 4
4 __ 4 __ 4 __ 4 = 5
4 __ 4 __ 4 __ 4 = 6
4 __ 4 __ 4 __ 4 = 7
4 __ 4 __ 4 __ 4 = 8
4 __ 4 __ 4 __ 4 = 9

There are many well-known variations to this puzzle . . . some of which have you use four 4s to construct expressions for all the numbers from 0 up through 100 (and even higher)! If you're interested, just search the Web for "four fours" and you'll find them. Enjoy!

INTRO TO EQUATIONS

But we haven't yet covered all that we can do with variables. Remember earlier in this chapter when we set our variable $\square = 2$, and then plugged it into the expression

$3 + \square$

to get

$3 + \boxed{2} = 5$

Well, that was all well and good and it made everything look exactly the same as it does in basic arithmetic (which is reassuring), but If you think about it you'll see that this last problem can be changed up a little to look like this instead

$3 + \square = 5$

Now, let's pretend that we don't remember ever setting $\square = 2$ to come up with the problem. In fact, let's pretend that we don't remember coming up with the problem ourselves at all. Instead, let's imagine that we stumbled upon the problem exactly as we see it here.

If you do that, you'll see that we're staring at an entirely new kind of problem which asks us to figure out what value a variable must have in order to make a statement true. In algebra, this type of statement typically goes by another name . . . and it's probably a name that you're familiar with—they're called equations. Even though this problem isn't particularly tough to solve (we don't really need algebra to see that $\square$ must equal 2), it shows what algebra is all about! By that I mean that it takes us from a fragmented world containing nothing but phrases, to a rich and brave new world full of sentences. In other words, it takes us from expressions to equations.

WHAT ARE EQUATIONS?

And with that we've arrived at probably the most important concept in all of algebra—equations. In many ways, equations aren't new to you. In fact, you've actually been using them since you learned basic arithmetic. As you'll recognize, a simple problem like $3+2$ is really just a mathematical expression. And way back in the day when you said that the answer to this problem was 5, you were effectively writing an equation that said $3+2=5$.

The best place to start our look at equations is by answering the question: What do equations mean? We'll make a quick pass at answering this question now, and then we'll come back to it in a bit for a more detailed answer. To get started, let's take a look at the simplest equation that I can think of

$$1 = 1$$

Yes, that's actually a legitimate equation since it has two expressions located on either side of an equals sign—that's all that an equation needs! So, the big question is what does it mean? Well, the quick and dirty answer is that the sole purpose of an equation is to boldly proclaim that the expression on the left side of the equals sign has the exact same value as the expression on the right side. So, for the case of $1=1$, this equation means that the number 1 has to be equal to the number 1. Which, of course, it most definitely is. And the same thing must be true for all equations . . . even those big, burly, mean-looking ones.

HOW TO WRITE AND SOLVE AN EQUATION

Now let's step up the complexity ladder one rung and look at an equation that's a bit less straightforward that $1=1$ (which isn't too hard to find!). Here's our new equation:

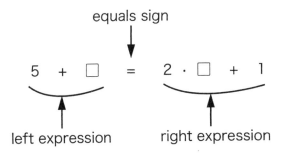

As you can see, this equation also contains two expressions located on the left and right sides of an equals sign . . . as all equations must. But this time, both expressions include numbers, variables, and operators. The two expressions look quite different, but the fact that they're on opposite sides of an equals sign means that they must both evaluate to the same value when you plug in the right number for $\square$. Of course, that means that you can't just plug in any old number for $\square$ . . . there must be a certain special value of $\square$ that does the trick.

So how can we find out what this special value is? Well, that special value we're looking for is called the solution to the equation, and the process of finding the solution is called solving the equation . . . and there are many ways to do it. The most basic way is to start plugging in numbers. By that I mean start with $\square = 1$, and calculate the values of the left and right expression. If they're equal, then you're done. If not, then move on and calculate $\square = 2$. Check again, and repeat as necessary. Here are the values of the left and right expressions of our equation for $\square = 1$ through $\square = 5$.

$\square$	$5 + \square$		$2 \cdot \square + 1$
1	6	>	3
2	7	>	5
3	8	>	7
4	9	=	9
5	10	<	11

As you can see, for $\square = 1$, 2, and 3, the value of the left expression is greater than the value of the right expression. And for $\square = 5$, the value of the left expression is less than the value of the right expression. And lo and behold, right there at $\square = 4$, we see that both the left and right expressions have the exact same value of 9. Which means that our equation has the solution $\square = 9$.

Okay, while it's wonderful that we solved this problem, it's pretty clear that this kind of plugging in numbers approach to solving algebra problems isn't going to cut it in the long run. After all, what if the solution was a huge number . . . it would've taken us forever to have gotten to it. So, we need to do better. And that's precisely where the methods of algebra that people have developed over the past many centuries, and that we're going to learn about, come to the rescue. Yes, there are *much* more efficient ways to solve problems like this . . . and much more complex problems to solve too.

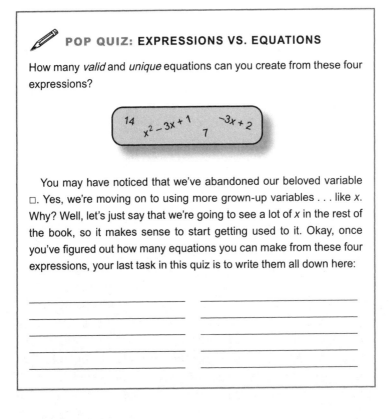

POP QUIZ: EXPRESSIONS VS. EQUATIONS

How many *valid* and *unique* equations can you create from these four expressions?

$$14 \qquad x^2 - 3x + 1 \qquad 7 \qquad {}^{-}3x + 2$$

You may have noticed that we've abandoned our beloved variable □. Yes, we're moving on to using more grown-up variables . . . like *x*. Why? Well, let's just say that we're going to see a lot of *x* in the rest of the book, so it makes sense to start getting used to it. Okay, once you've figured out how many equations you can make from these four expressions, your last task in this quiz is to write them all down here:

_____   _____
_____   _____
_____   _____
_____   _____

HALFTIME RECAP

Okay, before we continue on and really dig into understanding the meaning of equations, let's take a minute to recap what we've learned so far in this chapter. To do that, let's pull everything together in the context of an algebraic equation that we're by now quite familiar with—that's right, once again it's the Pythagorean theorem. I probably don't need to remind you, but the theorem can be written in the form of an equation that looks like

$$a^2 + b^2 = c^2$$

And now, for the first time, we really understand what it means to call this an equation. Here's a rundown of all the terms that we've covered so far in this chapter, and a summary of how they relate to the Pythagorean theorem.

- **Variables:** a, b, and c are all variables
- **Operators:** $+$ is an arithmetic operator
- **Expressions:** a^2+b^2 and c^2 are both examples of mathematical expressions since they are combinations of numbers, variables, and operators
- **Equations:** the whole thing, $a^2+b^2=c^2$, is an equation since it contains two expressions on either side of an equals sign

As we've now come to understand, the point of this famous equation is to say that after we plug in actual numerical values for the variables a, b, and c, the combination of variables in the expression on the left side of the equals sign must have the same numerical value as the combination of variables in the expression on the right. At least, that's the algebraic way of looking at things. And though this algebraic view is fantastic and helpful, it's good to keep in mind that it's not the *only* possible view. For example, this may surprise you, but we can picture what the Pythagorean theorem means using pictures too. In fact, let's do exactly that . . .

| | | | M | A | T | H | | B | R | A | I | N | | G | A | M | E | | | |

HOW TO PICTURE THE MEANING OF ALGEBRA

Take a look at the following drawing of the classic 3–4–5 right triangle that we became well acquainted with in the last chapter (the triangle is the dark gray bit in the middle).

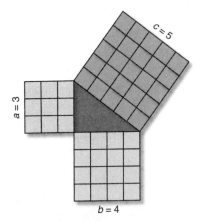

Remember that the 3–4–5 here means that the two legs of the tri-angle are 3 and 4 units long (in other words: $a=3$ and $b=4$). So, using $a^2+b^2=c^2$, we can therefore figure out that the hypotenuse must be

$$c = \sqrt{3^2 + 4^2} = 5$$

You can see the three lengths of the two legs and hypotenuse of the triangle shown on the drawing above. But why did I also draw all of those extra little squares? Here's a hint: You can use those little squares to help interpret the meaning of the algebraic equation $a^2+b^2=c^2$. In other words, those little squares can help us to picture the meaning of the Pythagorean theorem! How? Well, before mov-ing on, I'm going to let you think about that for a minute.

THE GEOMETRIC MEANING OF EXPRESSIONS

Okay, are you ready for the answer? Let's start by taking another look at the expression on the left hand side of the Pythagorean theorem:

$$a^2+b^2$$

One way to think of this expression—the algebraic way—is as something that says: "Hey, buddy! Give me two numbers—then I'll go ahead and square them both, add those results together, and finally give you back the total. Sound good?"

But, as I said before, the algebraic way of thinking about the Pythagorean theorem isn't the only way. What does that mean? Well, take a look at the single square at the top left of this drawing.

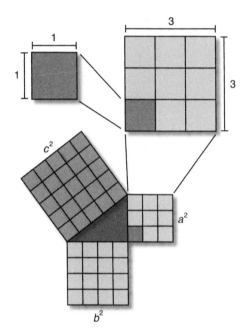

The sides of that little square are each one unit long. Now, if you look to the right of that square, you'll see a larger square whose sides are each three units long. The first thing I want you to think about is how many of the smaller squares can fit inside the larger square?

Well, as you can see in the drawing, you can fit three of the smaller squares from left to right across the larger square, and you can also fit three of the smaller squares from top to bottom down the larger square. So that means a total of 3 • 3 = 9 of the smaller 1-by-1 unit squares will fit inside the larger 3-by-3 unit square. Now imagine that the larger square didn't have a length and width of 3 small squares, but instead had a length and width represented by the variable *a*. In other words, imagine that the larger square measures a smaller squares high and a smaller squares across. In that case, the number of smaller squares that will fit inside this larger square is equal to $a \bullet a = a^2$.

Believe it or not, this means that our picture of the meaning of the left side of the Pythagorean theorem is complete. It says that you can think of the a^2 part of the equation as the number of small 1-by-1 boxes that fit inside a 3-by-3 box, and the b^2 part of the equation as the number of small 1-by-1 boxes that fit inside a 4-by-4 box. And c^2? Yes, you guessed it. You can think of it in the exact same way, but this time using the biggest box drawn whose sides are equal in length to the hypotenuse of the triangle.

THE GEOMETRIC MEANING OF EQUATIONS

Okay, those are nice ways to think about the meaning of a^2, b^2, and c^2 separately, but we're really interested in how you can put them all together to understand the entire equation. So, let's go ahead and combine the two boxes that contain a^2 and b^2 small squares. In other words, let's take all the 1-by-1 boxes that make up the 3-by-3 and 4-by-4 squares and combine them together into a bigger square, like you see in this drawing:

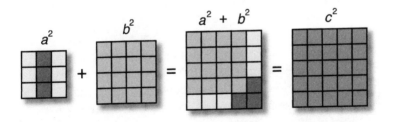

What does this show us? Well, we see that the big box created by adding a^2 of the 1-by-1 boxes to b^2 of the 1-by-1 boxes—which is a total of $a^2 + b^2$ of the 1-by-1 boxes—is the exact same size as the big box on the right that contains c^2 of the 1-by-1 boxes. Which means that $a^2 + b^2 = c^2$ works in this way for a 3–4–5 triangle.

Of course, we already knew that the equation for the Pythagorean theorem worked algebraically. In other words, we knew that we could plug in numbers and check that the equation was true. But now we have another way to think about what the equation means . . . we can think of it geometrically in terms of combining and counting boxes. Problems in algebra can often be looked at in more than one way, and looking at them from these alternate vantage points is a great way to understand what algebra really means.

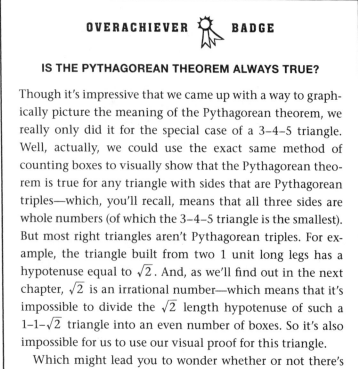

OVERACHIEVER BADGE

IS THE PYTHAGOREAN THEOREM ALWAYS TRUE?

Though it's impressive that we came up with a way to graphically picture the meaning of the Pythagorean theorem, we really only did it for the special case of a 3–4–5 triangle. Well, actually, we could use the exact same method of counting boxes to visually show that the Pythagorean theorem is true for any triangle with sides that are Pythagorean triples—which, you'll recall, means that all three sides are whole numbers (of which the 3–4–5 triangle is the smallest). But most right triangles aren't Pythagorean triples. For example, the triangle built from two 1 unit long legs has a hypotenuse equal to $\sqrt{2}$. And, as we'll find out in the next chapter, $\sqrt{2}$ is an irrational number—which means that it's impossible to divide the $\sqrt{2}$ length hypotenuse of such a 1–1–$\sqrt{2}$ triangle into an even number of boxes. So it's also impossible for us to use our visual proof for this triangle.

Which might lead you to wonder whether or not there's some other way to visually prove that the Pythagorean

Cont'd

Cont'd from page 104

theorem is true for *all* right triangles . . . no matter what the lengths of their sides? Well, take a look at this:

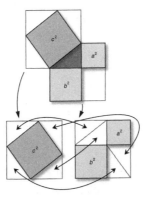

Do you see how this drawing proves the Pythagorean theorem? I don't blame you if you don't see it right away, so let's take a minute to walk through what's going on. First, notice that I've drawn a picture showing the squares a^2, b^2, and c^2 extending out from the sides of the darkly shaded right triangle near the top . . . just like we used in our earlier visual proof for the 3–4–5 triangle. Next, I've drawn two boxes that are each the exact same size surrounding these squares—one box is around a^2 and b^2, and the other one is around c^2. Now, here comes the big step: I've then redrawn each of these two big boxes—the one on the left contains the square c^2, and the one on the right contains the squares a^2 and b^2.

And, all of a sudden, something awesome has appeared. What's that? Well, the four curvy lines with arrows show that each of the four white triangles in the left box (the triangles formed in the regions between the outside box and c^2) fit perfectly into the halves of the two white rectangles in the right box. And since the two big exterior boxes are the

Cont'd

Cont'd from page 105

> exact same size, the fact that all of the white area on the left corresponds to all of the white area on the right means that the leftover area on the left (that is, the square c^2) has to be the exact same size as the leftover area on the right (that is, $a^2 + b^2$). So what does that mean? Well, it means we've proved that $a^2 + b^2 = c^2$ for all right triangles!

HOW YOU SHOULD THINK ABOUT EQUATIONS

Okay, it's time to dive in deep and really figure out exactly what equations mean. If you find it helpful to have a picture in your mind, then you can think of an equation as a scale that takes the stuff on the left and right sides of its equals sign and checks to see if they balance. In the case drawn below, the scale is balancing the two sides of the Pythagorean theorem.

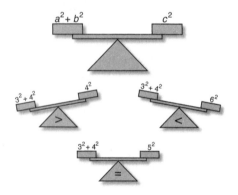

When the two sides of the scale are balanced, the equality is true. But when one side is larger than the other, the scale falls out of balance and the equation is not true. This is actually a rather powerful metaphor because it not only gives us a way to think about what an equation means, but it also shows us what we can and cannot do to an equation to solve it. We'll talk about this in more detail in a minute, but the basic idea is that we can do anything we want to an equation as long as we don't throw it out of balance. As we've seen, if we throw the equation out of balance, the equality will no longer be true, and the equation will therefore become impossible to solve.

HOW TO SOLVE AN EQUATION

The time has come for us to walk step by step through the process of solving algebraic equations. Haven't we done that already? Well, yes . . . sort of. But in truth, we've really only tiptoed around solving equations so far. Which means that it's now time for us to tackle the problem head-on and come up with a solid plan of attack for you to use in the future.

Since we've become so familiar with the Pythagorean theorem, let's use that famous equation in this tutorial. And, since we've also become so familiar with the art of building beautiful decks, it only makes sense that we're once again going to don our tool belts and hard hats. But first, we need to learn about the golden rule of equation solving.

THE GOLDEN RULE OF SOLVING EQUATIONS

Herein lies the golden rule of equation solving. Learn it, live it, love it, follow it, and all will be well. Ignore it, and all will be lost. Seriously . . . you want to follow this rule! But enough admonishment, here's the rule:

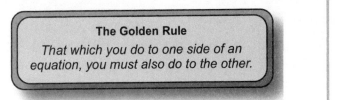

The Golden Rule

That which you do to one side of an equation, you must also do to the other.

◀ Q&DT

Okay, that sounds simple enough, but let's look at a few examples to make sure we fully understand all the ways to apply it. To help us, let's once again bring out our Pythagorean scale—complete with equivalent $a^2 + b^2$ and c^2 weights on its left and right sides.

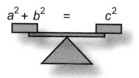

The basic idea conveyed by the golden rule is this: Whenever you do something to one side of an equation that changes its value (or the weight on that side of the scale), you must do the same thing to the other side of the equation (or to the weight on the other side of the scale) to ensure that its value changes by the exact same amount. If you don't, then the two sides of the equation will be thrown out of balance, and the scale will tip to one side or the other. Here are some specific examples of how to apply the golden rule . . . make sure you can fill in the blanks of all the expressions on the right to keep the equations balanced.

- **Addition:** If you add an expression (it can be a number, a variable, or some more complicated combination of numbers and variables) to one side of an equation, you must add the same expression to the other side.

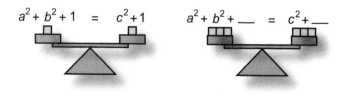

- **Subtraction:** If you subtract an expression from one side of an equation, you must subtract the same expression from the other side.

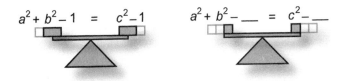

- **Multiplication:** If you multiply one side of an equation by an expression, you must multiply the other side by the same expression.

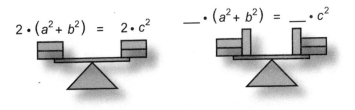

- **Division:** If you divide one side of an equation by an expression, you must divide the other side by the same expression.

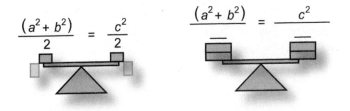

- **Square Roots:** If you take the square root of one side of an equation, you must take the square root of the other side too.

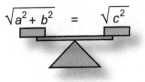

- **Exponentiation:** If you raise one side of an equation to a power (such as squaring it), you must raise the other side to the same power.

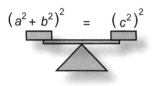

$$\left(a^2 + b^2\right)^2 = \left(c^2\right)^2$$

The take-home message here is simple: Don't do anything that unbalances an equation. Why? Because if you do, you'll destroy the equation! In other words, if you don't treat the two sides of an equation equally, then the stuff on the left side won't equal the stuff on the right side anymore—which means that it's not an equation anymore! But there's one thing I haven't mentioned yet . . . and that's why would we ever want to do any of this stuff to equations in the first place? Well, that's what the rest of this tutorial is about.

A CHALLENGE OF PYTHAGOREAN PROPORTIONS!

As rumor has it, a group of rival deck builders has proclaimed that any rookie deck builder can construct a *square* deck with perfect right angle corners, but only a true master can create a *rectangular* deck with those perfect right angle corners.

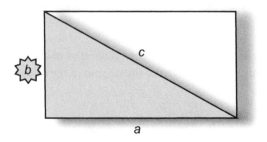

Apparently this group of deck builders has also scoffed at our method of calculating and then measuring out the length of the diagonal across the deck as a means of ensuring it ends up square. They've been overheard saying: "That's too easy . . . the equation is almost solved for you!" And just today, they issued you a direct challenge: If you're asked to build a rect-angular deck with a long side that's 15 feet long and a diago-nal that's 17 feet long, can you figure out how long the other side has to be to ensure that the deck has square corners? Of course you can—you just need to figure out how to solve for the variable b in the Pythagorean equation: $a^2 + b^2 = c^2$. So, how do we do it?

■ STEP 1: SIMPLIFY EACH SIDE OF THE EQUATION

The first step in solving an equation is to make the equation as simple as possible. We'll return to the Pythagorean equa-tion in a minute, but first to see what I mean by "simple," let's take a look at a different equation that needs a lot of simpli-fying:

$$2 + x - 2 \bullet 5 = 4 \, / \, 2 - x$$

There are several parts of the expressions on both the left and right sides of this equation that we can simplify accord-ing to the rules laid out by the order of operations (aka, PEM-DAS). Let's start with the left side expression, $2 + x - 2 \bullet 5$. First, we can perform the multiplication to simplify this to

$$2 + x - 10$$

Now we can swap the order of the first two parts of this expression to get $x + 2 - 10$, and then perform the subtraction. The left side expression therefore simplifies to

$$x - 8$$

That's a lot simpler, right? On the right side of the equation we can perform the division in $4 \, / \, 2 - x$ to get

$$2 - x$$

When we combine these two simplified expressions, we get a simplified form of the original equation that looks like

$$x - 8 = 2 - x$$

This new equation is equivalent to the first one (meaning we've been careful not to throw anything out of balance), but it's *much* simpler. In case you're wondering why we didn't use the Pythagorean equation as our example, well, as you can verify yourself, it's already as "simple" as it can get!

■ STEP 2: MOVE VARIABLE TO ONE SIDE (VIA ADDITION AND SUBTRACTION)

Okay, the next step in solving an equation for a variable is to move every part of the equation that contains the variable you're solving for to one side of the equals sign or the other. In our problem with the Pythagorean equation, b is on the left side of the equation, so let's focus on keeping it there and getting rid of everything else from that side of the equation. How can we do that? Well, we now know that we're free to subtract an expression from one side of the equation as long as we subtract the same expression from the other side of the equation too. So, let's get rid of the a^2 on the left side of the Pythagorean equation by subtracting it from both sides. Let's take a look at how this works in detail. First, subtract a^2 from both sides

$$a^2 + b^2 - a^2 = c^2 - a^2$$

Now, for the left side, swap the first two parts of the expression $a^2 + b^2 - a^2$ to get

$$b^2 + a^2 - a^2$$

which just simplifies to b^2.

The net result is that we've now turned the Pythagorean equation into

$$b^2 = c^2 - a^2$$

Remember, our goal for this step was to isolate the variable b on one side of the equals sign, and that's exactly what we've managed to do!

■ STEP 3: ISOLATE THE VARIABLE (VIA MULTIPLICATION, DIVISION, EXPONENTIATION, AND TAKING ROOTS)

The name of the game in the third step is to do whatever you have to do to get a single power of the variable you're solving for all by itself. The exact process you need to go through is different in every problem. Sometimes you'll need to use multiplication or division to get the variable by itself, sometimes you'll need to raise the variable to a power, and sometimes—as in our problem—you'll need to take a square root. If we take the square root of both sides of our equation, we get

$$\sqrt{b^2} = \sqrt{c^2 - a^2}$$

Of course, $\sqrt{b^2}$ is just equal to b (since taking the square root of a number undoes squaring it), which gives us

$$b = \pm\sqrt{c^2 - a^2}$$

Whoa! What's up with that "±" sign?

ALGEBRA DECODER! 🔍

• **PLUS OR MINUS: ±**

Remember that equations like $x^2 = $ "stuff" always have two solutions—one is equal to $+\sqrt{\text{"stuff."}}$ and the other is equal to $-\sqrt{\text{"stuff."}}$ For example, $x^2 = 4$ has the two possible solutions $x = \sqrt{4} = 2$ and $x = -\sqrt{4} = -2$ But instead of writing both solutions out this way, we can instead write the solution as $x = \pm 2$. The $\pm$ symbol is pronounced "plus or minus," so ± 2 is pronounced "plus or minus two."

So solving for b gives us two possible solutions:

$$b = \pm\sqrt{c^2 - a^2}$$

But, in this case, we're really only interested in one of the solutions—the positive one. Why? Well, since we're building a deck, we absolutely know that the solution must be a positive number . . . what would a negative length even mean? So for this problem, the single solution that we're after is

$$b = \sqrt{c^2 - a^2}$$

This method of using logic to eliminate a perfectly valid mathematical solution might seem sneaky to you. How can I just ignore something that math says is correct? The truth is that I'm not ignoring it—the negative solution is valid—it just doesn't make any sense for the particular problem we're looking at . . . since the problem we're looking at has to do with the real world where things can't have negative lengths!

Okay, as you'll recall, this whole problem started with a challenge to you that boiled down to seeing if you could figure out the value of b when $a = 15$ and $c = 17$. And you can now do it—you just need to plug the values of a and c into the equation we just found:

$$b = \sqrt{17^2 - 15^2} = \sqrt{289 - 225} = \sqrt{64}$$

Well that's an interesting solution . . . we've previously stumbled upon the fact that 64 is a perfect square—it's equal to 8^2. Which means that $\sqrt{64} = \sqrt{8^2} = 8$. That's interesting for three reasons: First, it tells us that the answer to our problem is 8. Second, it proves that you are indeed a master deck builder (congrats!). And third, it tells us that the particular triangle we've been studying is actually an 8–15–17 Pythagorean triple . . . just like the ones we looked at in the last chapter!

■ STEP 4: CHECK YOUR SOLUTION!

Yes, we now have a solution. But no, we're not actually done with the problem yet. And that's because you should *always* go back and check your solution. In other words, take the value you get when solving for your variable, and plug it back into the original equation. If you did everything correctly, then that original equation must balance . . . meaning its left-hand side (often called LHS) and its right-hand side (often called RHS) must have the same value. So, let's try it and see:

LHS: $a^2 + b^2 = 15^2 + 8^2 = 289$
RHS: $c^2 = 17^2 = 289$

Success . . . both sides have the same value, which means that our answer must be correct! To sum up, when you're solving equations, just remember these four steps:

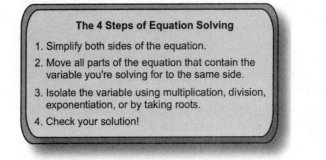

The 4 Steps of Equation Solving

1. Simplify both sides of the equation.
2. Move all parts of the equation that contain the variable you're solving for to the same side.
3. Isolate the variable using multiplication, division, exponentiation, or by taking roots.
4. Check your solution!

◀ Q&DT

POP QUIZ: LET'S SOLVE SOME EQUATIONS!

Every equation is different, so the only way to get good at solving them is by practicing. With that in mind, here's a quiz that has several equations for you to solve. You'll notice that each requires a slightly different approach (particularly when it comes to step three of our four-step solution method). Okay, here are your equations to solve:

1. $2 + x - 2 \bullet 5 = 4 / 2 - x$

2. $x / 3 + 3 = x / 2$

3. $\sqrt{x} - 4 = -\sqrt{x}$

4. $2x^2 - 3 + x^2 = 9 + x^2$

If you get stuck, take a look at the Math Dude's Solutions at the end of the book.

OVERACHIEVER  BADGE

HOW TO SOLVE AN EQUATION . . . GRAPHICALLY

There's no denying that equations like the ones we just solved can look pretty out there, and abstract. But there's a trick you can use to bring them to life and help you see what they mean. Want to know how? If so, read on as we take a deeper look at one of those equations: $\sqrt{x} - 4 = -\sqrt{x}$. The first thing we need to do is use the graphing skills we developed in the last chapter to plot the expressions on the left and right sides of the equals sign on the same set of axes.

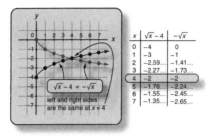

As you can see, the two curves representing the expressions from the left and right sides of the equation cross at the point $x = 4$. What do I mean by "cross"? I mean that the two curves have the exact same y values at that point—which means that $x = 4$ is the point at which both expressions are equal. And if both sides of the expression are equal there, then $x = 4$ must also be the point that makes the equation $\sqrt{x} - 4 = -\sqrt{x}$ true. So the solution to the equation must be $x = 4$, just as we found before . . . but now you know why! We'll look at some more examples of this way of looking at, thinking about, and solving equations in the next chapter.

- -
SECRET AGENT MATH-LIB
- -

WHERE ARE MY KEYS?

Good morning, secret agent _____ *(name of fruit or vegetable)*. Before you begin today's secret agent math mission, you need to find your _____ *(preferred mode of transportation)* keys! Each night before going to bed, you hide your keys in a different location. That way, if your identity is ever compromised and your house broken into, your _____ *(preferred mode of transportation)* will be safe! You remember hiding your keys in your _____ *(ridiculous place)* a few nights ago, but you're drawing a total blank about where you hid them last night. Fortunately, you've anticipated this problem and have prepared a secret code for you to decipher that will tell you precisely where your keys are hidden. All you have to do is crack it!

You take a peek at the encrypted message you left yourself and see:

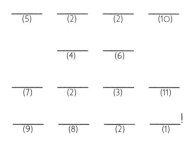

Wow, that looks kinda tough! You're struck with a momentary bout of panic (you really need those keys), but then you look down the page and see more—a whole lot more. Here's what it says:

> Good morning, me! This puzzle is all about the "order of operations." Remember that stuff? I sure hope so! Anyway, all you (meaning me) need to do is solve the following problems, and then plug the letters you get from the solutions into the secret message written above. After that, you'll know where to find your keys. Ready? Okay, here goes nothing...

PART 1—"WHERE ARE THE PARENTHESES?"

Each of the problems in this part look something like:

$2 + 3 \cdot 4 - 1 = 19$

Is this currently a valid equation? In other words, does the expression on the left have the same value as the one on the right? Well, using the order of operations (PEMDAS), we can find that the expression on the left is equal to

$$2 + (3 \cdot 4) - 1 = 2 + 12 - 1 = 14 - 1 = 13.$$

But the expression on the right side of the equation is 19, which means that this is not a valid equation...at least not yet. Because your job is to change the order of operations and turn it into a valid equation. How can you do that? By adding some well-placed parentheses.

So, where do those parentheses have to go to make $2 + 3 \cdot 4 - 1$ equal to 19? There are four possibilities that we'll consider, each of which will correspond to a different vowel that you'll need to help decode that secret message (you didn't forget about the message, did you?). Here are the options written out, along with the vowels corresponding to each of them:

(___ ? ___) ? ___ ? ___ .... ___E___

___ ? ___ ? (___ ? ___) .... ___I___

(___ ? ___ ? ___) ? ___ .... ___O___

___ ? (___ ? ___) ? ___ .... ___U___

The "?" symbols in these expressions indicate the positions of any of the big four arithmetic operators: $+$, $-$, $\cdot$, or $/$. So, which of these four possibilities is required for the problem $2 + 3 \cdot 4 - 1 = 19$ above? Is it E, I, O, or U? Take a second and give it a shot.

And the answer is...E. Do you see why? The parentheses for option E are located around the first two numbers, like this:

$$(2 + 3) \cdot 4 - 1$$

If we now follow the normal order of operations, we get $5 \cdot 4 - 1 = 20 - 1 = 19$. Which is indeed the answer we were looking for.

Okay, now that we're up to speed on how things are going to work, here are the four problems you need to solve. Remember, for each one, you need to figure out which pattern of parentheses from earlier is required to make the expression on the left of the equals sign have the same value as the number on the right. Make sense?

1. $(2+3) \cdot 4-1= 19$ ....____E____

2. $10+4+7 \cdot 2= 42$ ...._____

3. $3 \cdot 3+4 / 7= 3$ ...._____

4. $2 \cdot 6 \cdot 8-6= 24$ ...._____

PART 2—"WHAT ARE THE VARIABLES?"

To solve the next four problems, you need to solve the given equations for the variable x. For example, here's the first problem:

$$26-x \cdot 2+2=4$$

So, what's x here? Well, by now you're something of an expert at solving equations like this, so I'll leave it to you to use our four-step method and figure out that the solution is 12. Just to make sure I'm not crazy, let's go ahead and check this by plugging it back into the original problem:

$$26-12 \cdot 2+2=26-24+2=2+2=4$$

So that does indeed work since both expressions are equal to 4. Which means that the answer to this particular problem is 12. But wait, that's a number...we need letters to plug into our secret code. What letter should we use? Well, what's the twelfth letter in the alphabet? A, B, C,...it's L. Okay, we're now ready, so let's take a look at the problems:

5. $26 - x \cdot 2 + 2 = 4 \rightarrow x = \underline{12}$ .... <u>L</u>

6. $(2+3) / 5 + x = 15 \rightarrow x = \underline{}$ .... _____

7. $15 / 3 + x / 5 = 10 \rightarrow x = \underline{}$ .... _____

8. $x \cdot (6 + 3 - 6) = 24 \rightarrow x = \underline{}$ .... _____

PART 3—"WHAT ARE THE OPERATORS?"

To solve the final three problems, you need to figure out what operator is needed to make the expression on the left of the equals sign have the same value as the number on the right. For example, which of the big four arithmetic operators does this problem need?

$$10 \cdot 2 + 18 \underline{} 9 = 22$$

Well, let's look at the four options:

$$10 \cdot 2 + 18 \underline{+} 9 = 47$$
$$10 \cdot 2 + 18 \underline{-} 9 = 29$$
$$10 \cdot 2 + 18 \underline{\cdot} 9 = 182$$
$$10 \cdot 2 + 18 \underline{/} 9 = 22$$

So, we needed to divide. What letter for decoding our secret message does this answer correspond to? Here's the key:

$$+ \rightarrow K$$
$$- \rightarrow R$$
$$\cdot \rightarrow Z$$
$$/ \rightarrow S$$

And here are your final three problems:

9. $10 \cdot 2 + 18 \underline{/}_{(+,-,\cdot,/)} 9 = 22$ .... **_S_**

10. $(11 \underline{}_{(+,-,\cdot,/)} 8) \cdot 2 + 11 = 17$ .... _____

11. $7 - 5 \underline{}_{(+,-,\cdot,/)} 4 \cdot 2 = 10$ .... _____

Excellent—you've finished the problems! Now, dear friend, the only thing left to do is crack that code. If you look again at your secret message, you'll see a little number at the bottom of each blank space. Just write the letters of the alphabet from the solutions to problems 1 through 11 in the proper spaces, and your keys will be in your pocket in no time. Good luck!

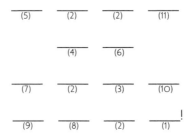

So, you set about solving the eleven problems, and then you decode your secret message. *(Make sure you've done it, because there's a spoiler coming!)* As soon as you see the message, you start laughing hysterically. Why? Because you've been wandering around all morning wondering why your right shoe felt so uncomfortable. You figured it must be a giant rock, but apparently it was just your _____ *(preferred mode of transportation)* keys. But hey, at least they were safe!

• ALGEBRA TUTORIAL •

HOW TO SOLVE AN ALGEBRA PROBLEM

It seems appropriate to finish off this chapter on the essentials of algebra with a tutorial that brings together everything we've learned so far. The goal here is to walk through the process of solving a real-world algebra problem from start to finish. So, with that in mind, let's get started . . . with a little story about cats and dogs.

A "REAL-WORLD" ALGEBRA DRAMA

Your dog loves toys. And you love giving them to him. Your cat, on the other hand, does not love your dog and therefore finds it amusing to hide his toys. You suspect that the cat is the culprit, so you begin to monitor his favorite hiding spot: the pile of towels next to his bed. But instead of constantly checking this spot for missing dog toys, you want to rig up an ingenious system that will automatically tell you how many toys are missing. How can you do it?

■ STEP 1: STOP AND THINK BEFORE DOING ANYTHING

The most important thing to do when you start working on a problem is to stop working on it. That might seem to make absolutely no sense, but the biggest mistake people make when solving problems is to start solving them too soon.

Instead, stop for a minute and think about what you need to do. Make sure you understand *exactly* what the question is asking and make sure you understand *exactly* what you are trying to solve for. So, let's follow this advice and stop and think about our problem for a minute.

First of all, could we even build something that would do this? Sure. Here's one way to do it. Put the pile of towels on a scale . . . or better yet, put them on a really smart scale that can send messages to your computer (yes, they exist!). That way, whenever your scale senses a weight increase, it can tell your computer that another toy has been hidden. Your computer can then use the equation we're going to come up with to figure out exactly how many toys are hidden. When that number goes above a certain limit, the computer can sound an alarm to let you know that it's time to go fetch!

■ STEP 2: ENGLISH TO EQUATION TRANSLATION

The second big step in solving an algebra problem is to perform an English to equation translation. In other words, we need to turn the words describing what our equation needs to do into an actual equation. In this case, we need to figure out how to write an equation that takes the current weight on the scale and gives us back the number of hidden dog toys. First, let's take the total weight on the scale, which we'll call W_{total}, and subtract the weight of just the towels, which we'll call W_{towels}. The difference between these two weights must be equal to the combined weight of all the dog toys, W_{toys}:

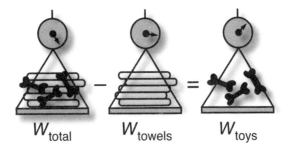

$$W_{total} \quad W_{towels} \quad W_{toys}$$

Which means we can write an equation that gives us the total weight of all the toys:

$$W_{toys} = W_{total} - W_{towels}$$

But we don't want to know the weight of the toys, we want to know how many there are. How can we do that? Well, if we know the total weight of all the toys, W_{toys}, and we divide that by the weight of a single toy, W_{toy}, then that gives us the total number of toys, N_{toys}:

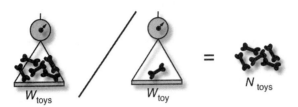

You should think about this for a minute and make sure you understand why it works. And once you do, you'll see that we can also write it in the form of an equation, like this:

$$N_{toys} = W_{toys} / W_{toy}$$

But how do we know the values of W_{towels} and W_{toy}? Well, we know because we were clever enough to measure them and write them down before we put the towels on our scale . . . or if we didn't, we'd better do that now!

■ STEP 3: SOLVE THE EQUATION FOR THE VARIABLE YOU'RE INTERESTED IN

The third step in solving an algebra problem is to solve the equation (or possibly the equations) you've come up with for the variable you're interested in. This step will usually entail going through the procedure outlined in the How to Solve an Equation tutorial. In this problem we want to solve for the total number of dog toys on the scale, N_{toys}. And if you look at the second equation we wrote down in our English to equation translation, you'll see that we have just such an equation:

$$N_{toys} = W_{toys} / W_{toy}$$

So now that we know that, we can go ahead and plug in numbers for the variables on the right side of the equation to figure out the total number of toys . . . right? Well, not so fast. We don't actually know the values of all the variables on the right side of the equation. In particular, we don't know the weight of all the toys, W_{toys} . . . or do we? Actually, if you look back up to the first equation in our translation, you'll see that we actually do. It's just

$$W_{toys} = W_{total} - W_{towels}$$

Which means that we can combine these two equations by plugging the second one into the first

$$N_{toys} = (W_{total} - W_{towels}) / W_{toy}$$

and the result is a new equation that gives us the total number of missing toys, N_{toys}, in terms of things that we know the values of: W_{total} (which is just what the scale tells us), W_{towels}, and W_{toy} (both of which we measured before). And this equation is *exactly* what we're after.

At this point in the process, it's time to take the abstract solution we've come up with (abstract in the sense of being written in terms of a bunch of variables), and turn it into a numerical solution. And to do that, all we have to do is plug in numerical values for all the variables on the right side of the equals sign. This step won't always be necessary. Sometimes you'll just want to solve a problem and leave it in terms of variables. For example, you'll notice that we don't actually have any numerical values to use for this problem! And that makes sense because our equation was only intended to be used to tell our computer how to convert from total weight on the scale to total number of dog toys. We weren't looking for a specific answer to a specific problem.

But what if we were? In fact, let's imagine that when we check the scale, we see that it has a total of five pounds of weight on it. If we also know that the pile of towels weighs two pounds and that each dog toy weighs one-half pounds (heavy toys . . . strong cat!), then we can use all these numbers and our equation to figure out how many dog toys are hidden:

$$N_{toys} = (5 - 2)/\tfrac{1}{2} = 3/\tfrac{1}{2} = 6$$

Which tells us that there are a total of six toys hidden at the moment . . . poor puppy!

■ STEP 4: MAKE SURE YOU UNDERSTAND THE RESULT

The fourth step in the algebra problem-solving process is closely related to the first. The name of the game here is to slow down and take a minute to think about the result you got. Ask yourself if it makes sense. If the answer is a negative number, you should ask yourself if you expected that. If it's a huge number, you should ask yourself if you expected that? The bottom line is this: Don't just declare that you're done because you've gotten a number out of the problem. The only reason you should declare that you're done is because you *understand* the number you've gotten out of the problem.

■ STEP 5: USE YOUR RESULT TO SOLVE OTHER PROBLEMS

The fifth and final step of the algebra problem-solving process is to use the result you've obtained to solve other problems. Why have I included this as a step? Because we're talking about solving real-world problems here . . . not textbook problems. In the real world, many problems that you solve will make you think of something else related that you need to solve too. And the good news is that this process is usually easier because you've already gone through the process and have a solution that you can use to help you in the next step.

For example, imagine that soon after coming up with your equation for N_{toys}, you realize that instead of always having to calculate the total number of hidden toys to see if you need to go fetch them, you'd rather just calculate the total weight of the scale when there were, say, four toys on it. That way, you'd only have to do the calculation once, and from that point on you could just check to see if the current total weight on the scale is over this critical amount. Well that just sounds brilliant! How can we do it? It's actually pretty easy—most of the work was done when you came up with the equation for N_{toys}.

What do I mean? Well, let's look at the equation for N_{toys} again:

$$N_{toys} = (W_{total} - W_{towels}) / W_{toy}$$

Given the total weight on the scale, this equation allows us to solve for the total number of hidden toys. But we could instead solve the equation for the total weight, W_{total}, so that given the total number of hidden toys, it will tell us the total weight. To do that, we just need to follow the steps outlined in the tutorial on solving equations:

Multiply both sides by W_{toy}: $N_{toys} \cdot W_{toy} = (W_{total} - W_{towels}) / W_{toy} \cdot W_{toy}$

Simplify the RHS: $N_{toys} \cdot W_{toy} = W_{total} - W_{towels}$

Add W_{towels} to both sides: $W_{towels} + N_{toys} \cdot W_{toy} = W_{towels} + W_{total} - W_{towels}$

Simplify the RHS: $W_{towels} + N_{toys} \cdot W_{toy} = W_{total}$

And that's the answer! If we flip the left and right sides of the last equation around to make it easier to read, we see that the total weight on the scale is given by

$$W_{total} = W_{towels} + N_{toys} \cdot W_{toy}$$

If you think about it, you'll see that this equation makes perfect sense. It just says that the total weight on the scale is equal to the weight of the towels plus the total weight of the dog toys (since the number of toys times the weight of one toy must be the total weight of all the toys).

Remember, our goal is to figure out what the total weight

will be when there are four toys hidden in the towels—which we'll call W_{fetch} since that's the weight when you need to go get the toys. And now we know everything we need to answer that question—since we already know that $W_{towels} = 2$ pounds and $W_{toy} = \frac{1}{2}$ pounds. So, let's plug these numbers in and find the answer:

$$W_{fetch} = 2 + 4 \cdot \frac{1}{2} = 2 + 2 = 4 \text{ pounds}$$

Which means that it's time to play fetch whenever that scale hits four pounds. And with that, we've finished our walk-through tutorial of a complete start-to-finish solution of a real-world algebra problem. Hopefully you're now a fan of the five-step plan!

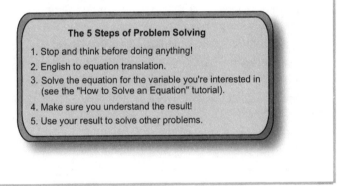

The 5 Steps of Problem Solving

1. Stop and think before doing anything!
2. English to equation translation.
3. Solve the equation for the variable you're interested in (see the "How to Solve an Equation" tutorial).
4. Make sure you understand the result!
5. Use your result to solve other problems.

WRAP-UP

That's it! Those are the essentials of algebra. From constants to variables, variables to expressions, and expressions to equations, you're now familiar with all the key ideas at the heart of algebra. And, in particular, you're now familiar with the fact that algebra is just arithmetic . . . with variables.

So what's left to explore? Well, think of algebra as a tree. So far we've investigated the trunk and some of the heftier branches. But there's a lot more to a tree than that. We still need to take a look at the smaller limbs, the leaves, and perhaps even some springtime buds. And after that, we'll be on our way to full algebraic bloom. Oh, come to think of it, we should probably take a look at the roots too . . . just to make sure the tree isn't going to fall over. So, before going any farther, let's pause, rewind, and check out those roots.

· FINAL EXAM ·

- ### THE BUILDING BLOCKS OF ALGEBRA: VARIABLES, EXPRESSIONS, AND EQUATIONS

Equations are made out of expressions. And expressions are made out of numbers, variables, and operators. Below you'll find a group of three left side expressions, three right side expressions, and three variable assignments that look like $x \rightarrow$ __ and mean "set x equal to such and such a value." Can you pick out the correct combination of variables and expressions to create three solved equations?

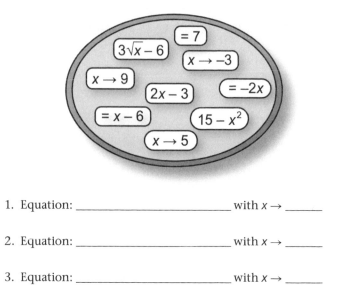

1. Equation: _____ with $x \rightarrow$ _____

2. Equation: _____ with $x \rightarrow$ _____

3. Equation: _____ with $x \rightarrow$ _____

• ORDER OF OPERATIONS

We had so much fun playing the Four Fours game earlier in the chapter that an encore seems appropriate. So here are five more numbers for you to try and create out of four number 4s, three operators, and as many sets of parentheses as you need:

4. 4 ____ 4 ____ 4 ____ 4 = 20

5. 4 ____ 4 ____ 4 ____ 4 = 24

6. 4 ____ 4 ____ 4 ____ 4 = 28

7. 4 ____ 4 ____ 4 ____ 4 = 32

8. 4 ____ 4 ____ 4 ____ 4 = 36

• SOLVING EQUATIONS

Back in the Expressions vs. Equations Pop Quiz, we constructed the following five equations from four expressions:

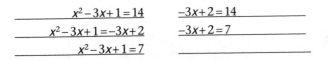

$$x^2 - 3x + 1 = 14 \qquad -3x + 2 = 14$$
$$x^2 - 3x + 1 = -3x + 2 \qquad -3x + 2 = 7$$
$$x^2 - 3x + 1 = 7$$

Now that we've learned how to solve equations, it makes sense to solve these too! But it's not as simple as that. Because there are two of these equations that we can't solve yet! So your assignment is to figure out which three equations you can solve, and then to solve them:

9. Equation #1: _____ Solution: $x =$ _____

10. Equation #2: _____ Solution: $x =$ _____

11. Equation #3: _____ Solution: $x =$ _____

• SOLVING ALGEBRA PROBLEMS

12. Let's return for a minute to the problem dealing with cats and dogs from the end of the chapter. At the end of that tutorial, we found an equation that gives the total weight on the scale, W_{total}, for various numbers of hidden toys, N_{toys}:

$$W_{total} = W_{towels} + N_{toys} \cdot W_{toy}$$

In the tutorial, we calculated the value of W_{total} for four hidden toys (that is, for $N_{toys} = 4$). But it sure would be nice to be able to see what W_{total} is for *any* value of N_{toys}, don't you think?

In fact, it's such a good idea that you should make a plot like the type we talked about at the end of the last chapter that shows this relationship. Specifically, you should calculate W_{total} for several different values of N_{toys}. To do this, you'll need to use the fact that $W_{towels} = 2$ pounds and $W_{toy} = \frac{1}{2}$ pounds. Each time you calculate W_{total} for a given value of N_{toys}, you will have found an ordered pair (N_{toys}, W_{total}) that you can plot on a graph. So, you should do exactly that—create a set of axes with N_{toys} on the *x*-axis, and W_{total} on the *y*-axis. Once you've plotted enough points to get a feel for the overall trend, you should draw a line that illustrates this trend.

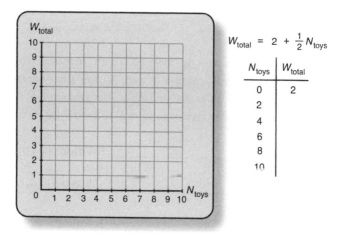

$$W_{total} = 2 + \frac{1}{2} N_{toys}$$

N_{toys}	W_{total}
0	2
2	
4	
6	
8	
10	

UNDERSTANDING
ALGEBRA BETTER

Where in each chapter we pause, rewind back to the roots of math, and then move forward toward a deeper understanding of algebra.

We learned to be happy. We danced 'round the hall. And learning to count was the key to it all.

—*Sesame Street*

CHAPTER 3

Walk the Number Line

> Five out of four people have trouble with fractions.
>
> —Popular T-shirt slogan

Since I'm the Math Dude, I could spend all day talking about why numbers are great; but the bottom line is that they're just plain useful. And their roots go deeper beneath the great tree of algebra than anything else. So in this chapter we're going to momentarily push the pause button on our algebra story and return back to math basics to talk about numbers: where they come from, what kinds there are, how we use them, and how to work with them. We need this foundation before we can get to more complex algebra. Once we're done with that, we'll once again hit the play button and move on to looking at how all of this basic math can help make solving algebra problems easier.

As you start reading this chapter you might think you already know this stuff and be tempted to skip ahead, but don't! This section is here for a reason, and the reason isn't just to review key math concepts. See, when many people learn math, they are taught just the "rules" but not the reasoning behind them. Sure, that can help you solve the exact problem you're being taught right at that second, but it doesn't give you the knowledge and understanding you truly need to solve the many other related math problems—problems you'll see on tests like the SAT and in higher-level math classes.

Continuing our theme of food-related metaphors that began in the introduction, being told you can't put aluminum foil in the microwave might prevent you from reheating that pizza in the foil it was wrapped in. But six months later it might not occur to you not to warm up your leftover spaghetti in the aluminum container it came in. Now, if you had initially been told *why* you shouldn't put aluminum foil in the microwave, you probably would've connected the dots. In other words, knowing that all aluminum (not just aluminum foil) + microwave = fire means that when faced with a new kind of aluminum—the kind your takeout container is made from—you'll remember what happens and you won't have to explain to your parents how you burned down the kitchen!

Not burning down the kitchen is, essentially, the point of the beginning of this chapter. Instead of relying on the "rules" you've been taught, I'll give you the deeper reasoning behind many familiar math concepts. And this new understanding will be your secret weapon for making sense of the algebra we encounter at the end of this chapter and throughout the rest of the book.

A BRIEF HISTORY OF NUMBERS

Let's kick off our look at numbers by traveling back in time. And when I say "back," I mean way back. How far exactly? About 30,000 years—back to the time from which bone fragments and other knickknacks with notches carved in them date to. People have speculated that the marks on these bones were used to help tally up and keep track of things like how many sheep they had or how many more days they needed to wait before harvesting their crops.

THE BIRTH OF MATH

But this system of counting with "tally-bones" is difficult to use at best. Why? Well, let's say you're making one notch on your tally-bone for every full moon since you planted some wheat. After four full moons, you look at your notches and see

||||

Which means that it's time to harvest your wheat! Okay, I'd say it actually looks like that system works pretty well. And that's true . . . in this case it is perfectly effective. But what if you were trying to keep track of the number of weeks that had passed instead of the number of months? Then your tally-bone would look something like

|||||||||||||||

It's now starting to get a bit more difficult to tell at a glance how long it's been, right? Even worse, what if you needed to keep track of the number of days that have passed—then you're looking at something like

And clearly now that's just ridiculous! Even if you didn't run out of room on your tally-bone, every time you wanted to know how many days had passed you'd have no choice but to start from the beginning and count each and every mark. That just doesn't seem very smart . . . there's gotta be a better way to do things.

EGYPTIAN ARITHMETIC

I'm not alone in thinking that notch-carving is an unwieldy system. After a while, everybody began to recognize that a more efficient method was needed. And about 5,000 years ago the Egyptians developed a system using hieroglyphics that was definitely an improvement. Here are the first five symbols in the Egyptian numeral system:

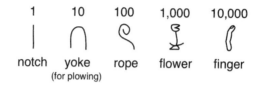

1	10	100	1,000	10,000
notch	yoke	rope	flower	finger
	(for plowing)			

Notice that the symbol for the number 1 is basically a notch—exactly like our tally-bone carving friends used. And, not accidentally, it looks a lot like our numeral "1." But unlike the tally-bone system, the Egyptians didn't stop with a single symbol. They also had symbols for the numbers 10, 100, 1,000, 10,000, 100,000 (a frog), and 1,000,000 (a human with raised arms). But what about the numbers 2, 3, and all the others that don't have specific hieroglyphics? Well, here's how they did it:

Do you see what they did? In order to write 2 instead of 1, they just wrote the symbol for 1 two times. But, you may say, that's exactly like the tally-bones! How is that more efficient? Well, the efficiency comes from the fact that they added those additional symbols we've seen for higher powers of 10. So while they did have to make an extra hieroglyph for each number between 2 and 9, when they got to 10, they went back to writing only a single symbol—which is definitely an improvement. But the Egyptian system is not without its flaws too (which is probably why we're not all using it today). The system may be *more* efficient than notch-carving, but that's not really saying much. After all, you still have to do quite a bit of work to write down a number like 3,358. So, once again it's clear that there's gotta be a better way. And, of course, once again there is: it's the ten digit "base-10 decimal system."

🖉 POP QUIZ: **ANCIENT ARITHMETIC**

Have you ever wished that you could travel back in time and visit ancient Egypt? It would be fun, but the math would take some getting used to. For example:

1. Think of a number greater than 1,000 but less than 10,000 _____
    Now translate this into Egyptian hieroglyphics:

2. Think of a smaller number that's still greater than 1,000 _____
    Now translate that into Egyptian hieroglyphics:

3. Add the two numbers together . . . Egyptian-style:
    _____ + _____ = _____
4. Now subtract the second number from the first . . . Egyptian-style:
    _____ − _____ = _____

How did that work out for you? Care to try multiplying or dividing this

Cont'd

Cont'd from page 139

> way? Are you now glad that you live in the modern-day world with our modern-day decimal system?

THE DECIMAL SYSTEM

The ten digits of the decimal system are, of course, the very same ones that you've known about since you first started having your fingers counted as a child:

1, 2, 3, 4, 5, 6, 7, 8, 9

But wait—that's only nine digits, right? Yep—so what's the tenth? Is it 10? Nope, that's just two digits used together. But which two? Aha! It's "1" and the elusive tenth digit . . . "0." There's been a long and complex history of people getting all riled-up over whether or not zero is really a number (since it represents "nothing") and what it means if it is. I like to picture ancient Greeks shouting, "How can nothing be something?" But no matter what your philosophy, zero is just plain useful . . . and it's absolutely necessary in the modern decimal system. Besides, without zero how could you say how many dollars you have left after you've spent your last one?

ALGEBRA DECODER!

• DIGITS

The system of numbers that you and I use every day needs only ten symbols (or digits) to represent all numbers. It's no coincidence that we use a ten digit system . . . just count the number of fingers on your hands and you'll know why. It's also no coincidence that the word "digit" doubles as the word for finger. Yes, people have been counting digits on their digits forever.

So, using our ten decimal digits we can count from 0 to 9 . . . but then what? Of course, you've known for years that the number after 9 is 10, but have you ever stopped to notice that the symbol for "10" is a combination of the digits 1 and 0. In other words, it's not an altogether different symbol! When you think about it, you'll see that this really is quite a clever idea. The number 10 is made up of one 10 and zero 1s. Similarly, the number 11 is made up of one 10 and one 1. And the value of 123 is made up of one 100, two 10s, and three 1s.

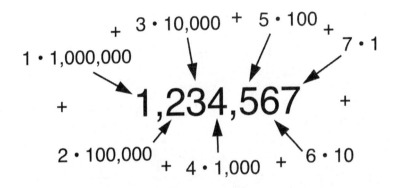

In other words, the farthest digit to the right represents how many 1s, the next digit to the left represents how many 10s, the next one how many 100s, and so on. Each successive place to the left represents one larger multiple of ten. Not only is that clever, it's also a lot more efficient than the ancient systems of writing numbers that we've looked at. Just imagine how much space it would take to write a number like 1,234,567 using notches in a tally-bone! Yeah, a whole lot. Which is exactly why the decimal system managed to become so popular.

ALGEBRA AND DECIMAL NUMBERS

Each of the digits in a number like 1,234,567 can be made by multiplying some number between 0 and 9 by a multiple of 10. For example, starting from the right side of this number, you can see that the ones digit is made from 7 • 1, the tens digit is made from 6 • 10, the hundreds digit is made from 5 • 100, and so on. Which means that *any* number in the decimal system can be written as

"any number" = . . . +n_{1000} • 1,000+n_{100} • 100+n_{10} • 10+n_1 • 1

In other words, we can think of any number as being made up of a bunch of variables which I've called n_1, n_{10}, n_{100}, n_{1000}, and on and on (some of these could equal to zero). So the variable n_{1000} contains the digit that appears in the thousands place, the variable n_{100} contains the digit that appears in the hundreds place, and so on. For the number 1,234,567, this means that $n_1 = 7$, $n_{10} = 6$, $n_{100} = 5$, $n_{1000} = 4$, and . . . well . . . you get the picture. So what? Well, this shows that even the most basic aspect of mathematics—numbers and counting—can be expressed with algebra!

POP QUIZ:
ALGEBRA, DECIMAL NUMBERS, AND ADDITION

Let's use this idea of writing numbers as the sum of a bunch of variables and multiples of 10 to figure out another (perhaps better) way to do addition, shall we? Here's what I have in mind:

I'll think of a number between 100 and 1000 .... _451_ (#1)
Now you think of another number between 100 and 1000 ...... _____ (#2)

Let's write each of these as the sum of three numbers times different multiples of 10:

#1: _451_ = _4_ • 100 + _5_ • 10 + _1_
#2: _____ = ___ • 100 + ___ • 10 + ___

Cont'd

Cont'd from page 142

> Now add up both numbers in each of the three columns for the hundreds, tens, and ones place. Write them below in the corresponding column.
>
> #1 + #2 = ___ • 100 + ___ • 10 + ___ = _____
>
> What do you think? Is adding numbers by powers of ten like this potentially faster than using the old-fashioned right-to-left method? Of course, once you're good at it, you don't have to write it all out like this . . . you can just do it in your head.

ALGEBRA AND THE NUMBER LINE

So now that we know something about the history of numbers, it's time that we start to figure out what kind there are, how they're related to each other, and how they all relate to algebra. And to start us off in that endeavor, let's take a look at a dear old friend . . . the number line:

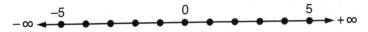

As you probably know, a number line is just a way for us to represent the continuum of numbers. They're typically drawn with arrows on both ends that inform you that if we had a magical sheet of paper, the line would extend out indefinitely in either direction toward what we call infinity (that's what those "∞" signs mean). As you can see, larger and larger numbers are shown heading off to the right in what's called the positive direction, and smaller and smaller numbers head off to the left in the negative direction. Which, of course, means that you can immediately tell the relative sizes of two numbers just by looking at their position on the number line—if one number is farther left than another, then that number must be smaller.

"Small" Numbers

It's easy to get tripped up when using the word "small" to describe numbers. For example, while it's true that a really tiny number like 0.00000001 is "small," it's still a lot bigger than a negative number like −1,000,000 . . . just look at the number line! But there's another meaning of the word "small" in which a number like 0.00000001 is actually much smaller than the number −1,000,000. What's the meaning? Well, it has something to do with what's called the "magnitude" of the numbers. But before we can talk about that, we first need to learn about absolute values.

ABSOLUTE VALUES

Okay, first things first: What's does "absolute value" mean? Well, the absolute value of a number is just its distance away from zero on the number line. That means that the absolute value of a positive number like 2 is just equal to 2 since that's how far away it is from zero on the number line.

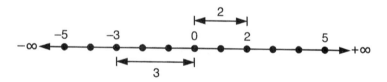

But things get a little more complicated when we talk about negative numbers. For example, what's the absolute value of the number −3? Well, how far away is the number −3 from 0? If you look at the number line, −3 is three "steps" away from 0. Which means that the absolute value of −3 is equal to 3. It doesn't matter if the steps were in the positive or negative direction, all that matters is the number of steps. In other words, the absolute value of a number only tells us about its size or "magnitude"—it doesn't care about its sign. And speaking of magnitude, we can now see that while the number 0.00000001 is greater than −1,000,000, its magnitude is actually much smaller. So, at least in one sense, 0.00000001 is indeed a "small" number.

ALGEBRA DECODER!

• **ABSOLUTE VALUE**

Okay, it's time to put a more algebraic spin on the idea of an absolute value. And to do that, we're going to need some notation. The absolute value of either a number or an expression in algebra is indicated by enclosing that number or expression between a pair of vertical bars. For example, we write the absolute value of the number −2 as $|-2|$ and the absolute value of the variable x as $|x|$.

HOW TO CALCULATE THE ABSOLUTE VALUE OF NUMBERS

The easiest way to find the absolute value of a lone number is to ignore any negative sign in front of it. So, $|-1| = 1$, $|5| = 5$, and so on. If you're instead looking to find the absolute value of an expression consisting of multiple numbers and operators, first simplify the expression, and then ignore any negative sign in front of the result. For example, the expression $|3 + 2 - 7|$ simplifies to $|-2|$, which is just equal to 2.

◄ Q&DT

HOW TO CALCULATE THE ABSOLUTE VALUE OF VARIABLES

So that's numbers, but how do you deal with absolute values when working with variables? Is it done the exact same way?

The absolute value of x is just x, and the absolute value of $-x$ is just x too, right? Well, not so fast. We don't actually know what x is . . . which means that x could end up being a negative number. And in that case, if you said that $|-x| = x$, then you'd say that the absolute value was a negative number—which certainly isn't right! So how does it work?

Well, let's think about how to write a general expression for the absolute value of the variable x. There are three scenarios that we have to consider:

- x will eventually represent a positive number
- x will eventually represent a negative number
- x will eventually represent the number 0

In the first case, when x is positive, the absolute value of x is just equal to the value of x. In the second case, when x is a negative number, we can find its absolute value just by changing it into a positive number—which is analogous to our "ignoring" the negative sign of ordinary numbers earlier. In the final case, when x is zero, we don't actually need to do anything to find its absolute value. Why? Because the absolute value of zero is just zero! If you think about these three cases for a minute, you'll see that we can put them all together and write an equation that allows us to find the absolute value of the variable x for any value that it might have:

Q&DT ▶

$$
\text{Absolute Value}
$$
$$
|x| = \begin{cases} x & \text{if } x > 0 \\ 0 & \text{if } x = 0 \\ -x & \text{if } x < 0 \end{cases}
$$

Do you see how this equation works for all three cases? The conditions on the far right tell you when you should use each line. In other words, you can read this equation like this:

- If $x>0$, then $|x|=x$, and you're done. If not, head to the next line . . .
- If $x=0$, then $|x|=0$, and you're done. If not, head to the next line . . .
- If $x<0$, then $|x|=-x$, and you're done!

But x here doesn't have to be just a single variable . . . it could actually be an entire expression! In other words, if you need to calculate the absolute value of something like $|z-1|$, you can still use the same equation (z here is just another variable). All you have to do is replace x in our earlier equation by the expression $z-1$, like this:

$$|z-1| = \begin{cases} z-1 & \text{if } z-1 > 0 \\ 0 & \text{if } z-1 = 0 \\ -(z-1) & \text{if } z-1 < 0 \end{cases}$$

In other words, if z is equal to 4, then $z-1=4-1=3$. Which means that we can use the top line of the equation to figure out the absolute value of $|z-1|$. Or, if $z=-4$, then $z-1=-4-1=-5$. And that means that this time we instead use the bottom line of the equation to find the absolute value of $|z-1|$. We'll talk more about equations like this a little later in this chapter.

POP QUIZ:
ABSOLUTE VALUES ON THE NUMBER LINE

Start by drawing a number line that extends from −5 to 5. Be sure to label the location of the origin (in other words, the point that's equal to zero) and some other numbers on the line so that you can find your way around. Draw your number line here:

Cont'd

Cont'd from page 147

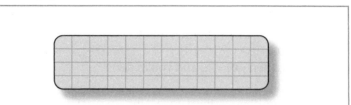

Now, mark the locations of the numbers −4, 3, 1, −2, 4, x, and y on your number line. You can put x and y wherever you want—they can be positive, negative, or zero. Afterward, calculate the following absolute values.

1. $|-4| = $ _____
2. $|3| = $ _____
3. $|-4+3| = $ _____
4. $|-4|+|3| = $ _____
5. $|-4-3| = $ _____
6. $|x+3| = $ _____
7. $|x-y| = $ _____

Looking at the points you have plotted on your number line, can you figure out what each of these values mean?

HOW TO CALCULATE THE DISTANCE BETWEEN NUMBERS

So we now know that the absolute value of a number or a variable is just the distance between that number and zero. But how do we find the distance between two numbers, two variables, or even between a number and a variable? Let's start by figuring out how to find the distance between the numbers 2 and −5. We know that the distances from 2 to 0 and from −5 to 0 are each given by the absolute values $|2|=2$ and $|-5|=5$. And, if we look at where 2 and −5 sit on the number line we can see that the distance between them is equal to the distance from −5 to 0 plus the distance from 0 to 2. In other words, it's equal to $|-5|+|2|=5+2=7$.

But is it always true that the distance between two numbers is the sum of their absolute values? For example, what if we want to find the distance between the numbers 2 and 5 instead of 2 and −5. Can we just add the distances from 0 to 2 and from 0 to 5 like $|2|+|5|=7$? No! As you can easily see by looking at the number line below, the distance from 2 to 5 isn't 7 . . . it's 3!

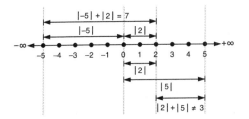

So what went wrong? And what's the right way to calculate the distance between two numbers? Well, let's think about what the sum of the absolute values of two numbers like we've been calculating really means. The absolute value of one number is its distance from zero, so the sum of the absolute values of two numbers is just the total distance from zero for both numbers together. When one number is positive and the other is negative, this is the same thing as the total distance between the two numbers. But, if you think about it, you'll see that when both of the numbers are either positive or negative, it does *not* tell us the distance between the numbers. But something else does . . . instead of the sum of the absolute values, let's look at the absolute value of the difference between the two numbers.

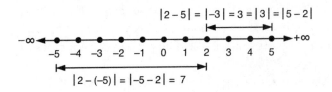

As you can see, $|5-2|=3$ . . . which is exactly the distance between 2 and 5! And it doesn't matter in which order we wrote the numbers—we could have written $|2-5|=|-3|=3$ and we still would have found the distance between 2 and 5. You can also see on the number line earlier that $|2-(-5)|=|2+5|=7 \ldots$ which is precisely the distance between 2 and –5! Again, the order of the numbers doesn't matter—we could just as well have written $|-5-2|=|-7|=7$.

WATCH OUT!

An Absolute (Value) Mistake

Be careful when working with absolute values not to make the mistake of trying to write that $|-5-2|=|-5|-|2| \ldots$ because it's not true! As you should verify, $|-5-2|\neq|-5|-|2|$. So be careful!

WHAT'S THE SECRET TO A SAFE COMMUTE?

Nice to see you again, secret agent _____ *(name of fruit or vegetable)*. Now that you've found your keys, it's time to get to work. But you, being a(n) _____ *(adjective)* math secret agent, can't simply drive your _____ *(preferred mode of transportation)* to work like a "normal" person. After all, you need to be sure to throw any would-be followers off the trail. And so, despite living a mere nine blocks from work, your commute takes you a good fraction of the morning. But it's worth it. After all, you haven't been captured yet!

So what's today's route? Dunno! That's right, every morning you're sent a series of math problems to solve that give you that day's route. You wouldn't want your commute to be too predictable, right? At least that's what your boss tells you. Okay, enough of the small talk—let's get on with today's problems. Take your time in solving them, but not too much—you've got important secret agent math duties to attend to.

In the first six problems, your job is to figure out what the various expressions are equal to. Then, if the answer is a positive number, write **2** on the first blank line to the right of the problem. And if it's a negative number, write **1** on the first blank line. Finally, if the answer is an even number, write the letter **R** on the second blank line to the right of the problem. And if it's an odd number, write **L** on the second blank line. Now, have at it!

- $|-6| = \mathbf{6}$ .... $\underline{\;2\;}$ · $\underline{\;R\;}$
- $|9-3| = \underline{\quad}$ .... $\underline{\quad}$ · $\underline{\quad}$
- $|5-8| = \underline{\quad}$ .... $\underline{\quad}$ · $\underline{\quad}$
- $-3 \cdot |-4| = \underline{\quad}$ .... $\underline{\quad}$ · $\underline{\quad}$
- $-|-4|^2 = \underline{\quad}$ .... $\underline{\quad}$ · $\underline{\quad}$
- $(-|-4|)^2 = \underline{\quad}$ .... $\underline{\quad}$ · $\underline{\quad}$

Okay, the final four problems are a little different. This time you need to compare the two expressions to figure out which is larger. If the left expression is greater than the right expression, write a greater than symbol on the first blank line and the letter **R** on the second. If the left expression is less than the right expression, write a less than symbol on the first blank line and the letter **L** on the second. For example, in the first problem, the expression on the left simplifies to 3 and the expression on the right simplifies to 5. Since 3 is less than 5, we write a less than symbol and the letter **L**. Remember, if you're having trouble figuring out which number is smaller, just go back to thinking about your friend the number line. The farther left you go on the number line, the smaller the number. And the farther right you go, the bigger the number. Okay, good luck!

- $|-3| \;\underline{\;<\;}\; |-5|$ .... $\underline{\;2\;}$ · $\underline{\;L\;}$
- $|5| \;\underline{\quad}\; |-6|$ .... $\underline{\;2\;}$ · $\underline{\quad}$
- $-7 \;\underline{\quad}\; -6$ .... $\underline{\;2\;}$ · $\underline{\quad}$
- $-|-19| \;\underline{\quad}\; -3 \cdot |-6|$ .... $\underline{\;2\;}$ · $\underline{\quad}$

Excellent! Now that you have those problems solved, you're ready to map out your route for this morning's commute. How do you do that? It's easy. You just have to follow a set of directions that your boss has given you, and turn either right **(R)** or left **(L)** according to the answers you got. But before we start drawing out your route, let's make a big list of all the turns you're going to be making. Working from the first problem to the last, write the sequence of **R** and **L** values that you got as answers. The number before the **R** or **L** (either **1** or **2**) indicates how many **R** or **L** values you should write.

$$\underline{\quad R \quad} \quad \underline{\quad R \quad} \quad \underline{\qquad} \quad \underline{\qquad} \quad \underline{\qquad} \quad \underline{\qquad} \quad \underline{\qquad} \quad \underline{\qquad} \quad \underline{\qquad}$$
$$\;\;(1)\qquad\;(2)\qquad(3)\qquad(4)\qquad(5)\qquad(6)\qquad(7)\qquad(8)\qquad(9)$$

$$\underline{\qquad} \quad \underline{\quad L \quad} \quad \underline{\quad L \quad} \quad \underline{\qquad} \quad \underline{\qquad} \quad \underline{\qquad} \quad \underline{\qquad} \quad \underline{\qquad} \quad \underline{\qquad}$$
$$\;(10)\qquad(11)\qquad(12)\qquad(13)\qquad(14)\qquad(15)\qquad(16)\qquad(17)\qquad(18)$$

Okay, the numbers in parentheses tell you which turn the **R** or **L** is for. All you have to do now is use the following instructions to draw your route on the map. And be sure to take the indicated right or left turns. Now, make sure the front door is locked and the teakettle is off, and then get in your _____ *(preferred mode of transportation)* and go!

DIRECTIONS

- Drive 2 blocks up 1st Ave. and turn **right** . Drive 1 block and turn **right** . Drive 1 block, pull an evasive U-turn, drive 1 block, and finally turn _____. (3) Drive 1 block and turn _____. (4) Drive 2 blocks and turn _____. (5) You should be at the corner of 1st St. and 3rd Ave.

- Now, drive 1 block and turn _____. (6) Drive 2 blocks and turn _____. (7) Drive 1 block and turn _____. (8) Drive 1 block and turn _____. (9) Drive 1 block, pull an evasive U-turn, drive 1 block, and turn _____. (10) Drive 1 block and turn _____. (11) You should be at the corner of 1st St. and 5th Ave.

- Next, drive 2 blocks and turn _____. (12) Drive 2 blocks and turn _____. (13) Drive 1 block, pull an evasive U-turn, drive 2 blocks, pull another evasive

U-turn, drive 1 block, and finally turn _____. Drive 2 blocks and turn
_____. You should be at the corner of 1st St. and 7th Ave.
(15)

- Finally, drive 2 blocks and turn _____. Drive 2 blocks, pull an evasive U-turn,
  (16)
  drive 1 block, and turn _____. Drive 1 block and turn _____. Drive 1
  (17)   (18)
  block, pull an evasive U-turn, and then proceed 2 blocks until you arrive at work!

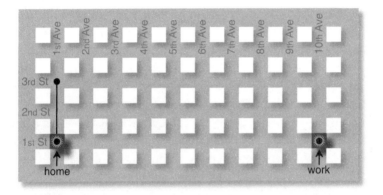

So, did you make it to work…before noon? It definitely wasn't
the fastest way to get to work, but it certainly was a safely circuitous route. After all, those U-turns are sure to have thrown off
any would-be followers. And, after studying the map, you realize
that your boss is even more clever than you'd ever imagined…
clearly she knows the secret to a secret agent's safe commute!

- -

HOW TO CALCULATE THE DISTANCE
BETWEEN VARIABLES

We now know how to calculate the distance between two numbers, but how do we extend this to calculate the distance between a variable and a number . . . or even two variables? The
good news is that it's done in exactly the same way. For example, on the following number line, the distance between the
number 4 and the variable x is equal to $|4 - x|$.

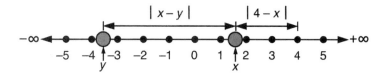

For $x=1$, the distance is $|4-1|=3$. Or for $x=2$, the distance is $|4-2|=2$. In the case of the drawing, the distance is somewhere between these two values. If we instead need to calculate the distance between the two variables x and y on the number line above, we simply need to calculate $|x-y|$. But, you might be thinking, why all this fuss about using the absolute value? In the number line above, we can see that $x>y$. Which means that $x-y$ is always going to be a positive number, and therefore that $x-y$ is equal to the distance between the two variables. Can't we just do away with the absolute value then? In one sense you're right, but in general we don't know in advance if $x>y$ or $y>x$. And if we don't know that, then we have no way of knowing if we should write the distance as $x-y$ (as we should if $x>y$) or $y-x$ (as we should if $y>x$). Which means that it's just a lot easier to use absolute values since they always ensure that the distance we calculate will be a positive number.

POP QUIZ:
WRITING DISTANCES ON THE NUMBER LINE

Using the following number line and the variables a, b, c, x, y, z . . .

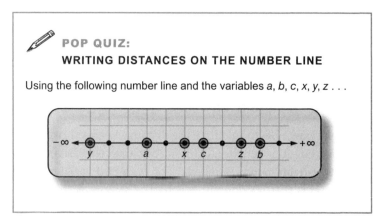

Cont'd

Cont'd from page 154

Complete these distance-related absolute value equations:

1. Write an equation that says that the distance between *b* and *c* is equal to the sum of two absolute values.

2. Write an equation that says that the distance between *z* and *a* is equal to the sum of three absolute values.

NUMBER BOOT CAMP

ZERO AND INFINITY

Contrary to popular belief, zero is not the smallest number and infinity is *not* the biggest. Why not? Well, as we've now seen, zero is the number with the smallest magnitude (or absolute value), but there are infinitely many smaller numbers on the negative side of the number line. And infinity is not the biggest number because you can always add 1 to the "biggest number" and then . . . oops . . . you have an even bigger number! The word "infinity" comes from the Latin *infinitas,* which means something like "endlessness." And that's a pretty good way to think about the meaning of infinity in math too. Infinity is not a number itself, but rather an expression of the unbounded nature of numbers. So when we say something like "there are an infinite number of this-or-thats," we are really saying that the things we're talking about don't end. Ever. Which does sort of boggle the mind, doesn't it?

POSITIVE AND NEGATIVE NUMBERS

As we talked about at the very beginning of the book, positive numbers are all the numbers greater than zero, and negative numbers are all the numbers less than zero. If you want a picture to have in your head, you can think of positive numbers as being like money that you have safely stored in your bank account.

And negative numbers? Well, they're like the debt that you owe the bank (or your parents). In other words, if you owe the bank $1,000 and you don't have any money in your account, then the current value of your account is −$1,000. That's right . . . it's worse than if you had nothing at all! And the next time you deposit $1,000 into the bank, the net value of your account will move from −$1,000 up to −$1,000+$1,000=$0. It's not much (or anything at all), but at least you'll be out of debt.

WATCH OUT!

Is Zero Positive or Negative?

If positive numbers are greater than zero and negative numbers are less than zero, where does that leave zero? Well, zero is special: It's the only number that is neither positive nor negative . . . which means that it has all the makings of a great trivia question.

INTEGERS

Any number (except zero!) can come in either a positive or negative flavor. But the most basic type of numbers that you've been using to count things your whole life are called integers. Here they are on a number line:

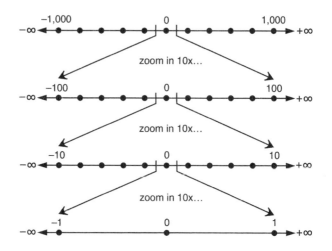

At the top you can see the number line extending from −1,000 to 1,000 (the tick marks represent intervals of 200). Below that we zoom in ten times to look a little more closely at the numbers ranging between −100 to 100 (the tick marks now represent intervals of 20). What do we see? Not surprisingly . . . more numbers. Below that we zoom in another ten times to look at the region of the number line from −10 to 10 (with tick marks spaced two apart), and so on.

Believe it or not, this sequence of number lines shows us something pretty remarkable about numbers. No matter how far you zoom in, there are always more of them. And if you think about it, you'll see that this is the same idea of infinity that we talked about before . . . but this time it's turned on its head. In other words, instead of saying that the number line goes on and on indefinitely to both the left and the right (which of course it does), we're now also saying that even just a tiny piece of the number line contains infinitely many numbers. The bottom line is that no matter which way you look, there are always more numbers.

RATIONAL NUMBERS

Or are there? I mean, if it's true that there are always more numbers, why is it that when we zoomed in the last time down to a number line extending from −1 to 1, there were only dots at −1, 0, and 1? Does that mean that there only three numbers in that range? Of course not! The problem is that we've been focusing on numbers without fractional parts . . . the integers. But there are plenty of other numbers out there besides integers. In fact, let's zoom in on the number line a little further to see what I mean.

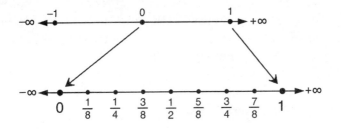

Fractions! Yes, they've all been here the whole time, we just couldn't see them. We've divided up the region between 0 and 1 into eight parts, but we could have divided the region into ten parts to make tenths, or fifty parts to make fiftieths, or any other number of parts to make a bunch of "whatever-eths" too.

Clearly these fractions are very different from integers—after all, integers don't have fractional parts. But these two types of numbers aren't completely unrelated. In fact, if you take another peek at those fractions written beneath the number line earlier, you'll notice that they are all made from two integers—one in the numerator (the top part) of the fraction and one in the denominator (the bottom part) of the fraction. Numbers like this that can be written as the ratio of two integers are called "rational" numbers (notice that the word "ratio" is part of the word "rational"). Which means that all fractions that are the ratio of two integers are rational numbers and, therefore, that all rational numbers can be written like a fraction:

$$\text{rational number} = m \,/\, n$$

What are m and n here? They're just symbols that can represent any integer values. For example, in the fraction $\frac{1}{2}$, $m = 1$ and $n = 2$; in the fraction $\frac{23}{32}$, $m = 23$ and $n = 32$; and so on.

Of course, we can also write fractions as decimal numbers: $\frac{1}{2} = 0.5$, $\frac{23}{32} = 0.71875$, and so on. You can easily convert a fraction into a decimal number simply by dividing the numerator of the fraction by its denominator. For example, to convert $\frac{23}{32}$ into a decimal number, all you have to do is divide 23 by 32 using a calculator (or using long division if you really enjoy that sort of thing). It's also true that any number that can be written as the ratio of two integers—in other words, any rational number—can be written in decimal form using a finite (meaning not infinite) number of decimal digits. Although, as in $\frac{1}{3} = 0.33333 \ldots = 0.\overline{3}$, one or more of those decimal digits may repeat forever (but since we know that it repeats, we don't need to keep writing it down forever).

If you think about it, you'll see that fractions written as the ratio of two integers aren't the only numbers that are rational. In fact, integers themselves are rational too. How's that? Well, it comes from the fact that any integer can be written as a fraction with a denominator equal to 1. For example, the number 3 can be written $^{3}/_{1}$ (think "three wholes") which means that it and every other integer is indeed a rational number. Now it's starting to look like all numbers are rational. Is that true? To answer this, let's zoom in on the number line one last time and take a look at the region between the fractions $^{3}/_{8}$ and $^{1}/_{2}$.

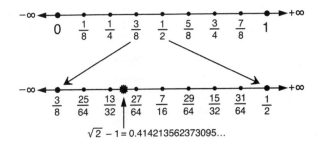

What do we find there? Of course, we find a lot more fractions: $^{25}/_{64}$, $^{13}/_{32}$, and so on. No matter how far we zoom in, we can always come up with more fractions. So does that mean that all numbers really are rational?

WATCH OUT!

Integer or Rational Number?

Is the number 5 an integer? Or is it a rational number? Actually, since we can write the integer 5 as 5 / 1, it's both! Don't get tripped up by thinking that it has to be one or the other. How about the number $12\frac{}{3}$? What is it? Well, it's definitely a rational number since it's written as the ratio of two integers. And you might notice that the fraction $12\frac{}{3}$ can be reduced to the number 4 which, of course, is an integer. So $12\frac{}{3}$ is a rational number, but it simplifies to have an integer value.

IRRATIONAL NUMBERS

Imagine you have a box containing ten pieces of paper labeled with the digits 0 to 9. Now imagine shaking this box around and pulling out one number at random. Write that number down, put it back in the box, shake it up, and then pull out another digit at random. Then keep on doing this . . . literally forever. If you could, the digits will also go on forever and, since they're random, they will never form a repeating pattern. Now, if you use this stream of digits to form a decimal number, you will have created what is called an "irrational" number. For example, imagine you randomly generate the digits: 1, 4, 1, 4, 2, 1, 3, 5, 6, and on and on forever (the forever part is the tricky bit), then you can use these digits to create the number:

$$1.41421356\ldots$$

And, as we talked about earlier, this number will be irrational since its pattern of digits never terminates or repeats (as it would if it were a rational number). What about the decimal point? Does it matter where we put it? No, think about it for a minute and you'll see that we could have put the decimal point between any two digits and the resulting number would still be irrational.

Why? Because regardless of where the decimal goes, the number still continues on for an infinite number of nonrepeating digits.

THE SQUARE ROOT OF 2

If you have a keen eye (and a little math background), you may have noticed that the series of numbers 1.41421356 . . . wasn't exactly pulled out of a hat. In fact, these are the first digits of a very famous irrational number: the square root of 2. Now, there are many famous irrational numbers including π (pi), ϕ (the golden mean), and e (Euler's number). But why, of all things, do people care about the square root of 2? That seems like kind of a random concern, right? Well, the answer goes back to the Pythagorean theorem (yes, once again) and the question: What's the length of the hypotenuse of a right triangle with legs that are each 1 unit long?

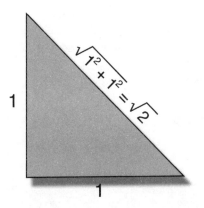

As we now know, we can use the Pythagorean theorem to find that the length of the hypotenuse of such a triangle has a length equal to $\sqrt{2}$. But way back in the day when people began trying to figure out the decimal equivalent of $\sqrt{2}$, they started to run into trouble. At first everything went along swimmingly. They found that $\sqrt{2}$ is a number that's a little bigger than 1, and then they found that it had to be a little bigger than 1.4, and then a little bigger than 1.41, and then (getting a little tired by now) a little bigger than 1.414, and on and on. No matter how many digits they calculated, the decimal digits just wouldn't stop coming. Why? Because $\sqrt{2}$ is irrational.

REAL AND IMAGINARY NUMBERS

In one sense all numbers are "real" (since I'm real, and I'm thinking about them). But as it turns out, only some numbers are *really* real. And that's because the word "real" has a particular meaning in math. Real numbers are all the numbers you can find on a real number line—which is the kind of number line that we've been looking at. That means that real numbers include all the integers, rational, and irrational numbers. But, you might be thinking, isn't that everything? Well, as it turns out, there's an altogether different type of number called an imaginary number that can be used to solve problems that real numbers can't deal with (such as the problem of solving for *x* in the equation $x^2 = -1$). Imaginary numbers can be combined with real numbers to create something even more complex called, sensibly, a complex number. You can picture the relationships between the various types of numbers like this:

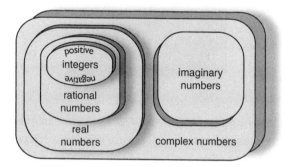

This picture says that every number is actually a complex number, but that there's a certain group of these numbers that are also called real numbers (those are the ones that we're working with). Within the realm of the real numbers, there is a certain group called rational numbers . . . and the rest are irrational. Finally, within the world of rational numbers, there is a group called integers . . . and the rest are fractions. It's pretty interesting to see that the world of real numbers that we're interested in is just a tiny part of the whole world of numbers. But the good news is that you don't need to worry about any of the more exotic types of numbers for now since we'll only be dealing with real numbers. Nonetheless, it's good to know that the others are out there. And besides, now you have something to look forward to learning about someday!

✏️ POP QUIZ: NAME THAT NUMBER

Can you figure out what type of numbers these are? Remember that a number can have more than one type. For example, looking at the figure earlier, we see than any rational number is also a real number . . . and a complex number too! For this quiz, you just need to give the most specific category: *integer*, *rational*, or *irrational*. But don't forget that everything in one of those categories is classified as a real number too!

1. 10 / 3 is a(n) .... _____.
2. $\sqrt{3}$ is a(n) .... _____.
3. −17 is a(n) .... _____.
4. $\sqrt{64}$ is a(n) .... _____.
5. π is a(n) .... _____.

And with that, you've completed math boot camp. You're now fully acquainted with all the numbers in the number zoo . . . as well as the line upon which they spend their time. Now that we're familiar with this first type of line—and in keeping with the theme of this chapter—let's move on to taking a look at some of the other roles that lines play in algebra.

LINEAR EQUATIONS

Q&DT ▸

What makes an equation "linear"? Well, the quick and dirty way to tell if an equation is linear is to take a look at the variable you're solving for . . . let's say it's called x. If the variable x shows up anywhere in the equation you're looking at with an exponent other than 1 (remember that $x^1 = x$)—for example, if it shows up looking like x^2, x^3, and so on—then the equation is *not* linear. Which means that linear equations can only contain constants and variables with exponents of 1 (which means no square roots of the variable). This might seem a bit arbitrary right now, but we'll see a little later in this chapter where it comes from.

✎ **POP QUIZ: LINEAR OR NOT?**

Are these equations linear? Make sure you simplify both the left and right sides of the equation as much as possible before you make up your mind.

1. $3 - x = x$ .... is linear? _____ (yes/no)
2. $5x^2 + 8 = 0$ .... is linear? _____ (yes/no)
3. $x \bullet x - 9 = 0$ .... is linear? _____ (yes/no)

WATCH OUT!

Linear Equation Imposters

Some equations look linear, but aren't. And others don't look linear, but are. Watch out for these imposters! For example, here's an equation that upon a quick glance might look linear:

$x + 1 = 1 / x$

Cont'd

Cont'd from page 164

But there's actually an x in the denominator of the right-hand side. And if you multiply both sides of the equation by x to get rid of that, we get the most definitely nonlinear equation

$x^2 + x = 1$

On the other hand, a problem like

$2 / x + 1 / x = 6$

doesn't look linear at all . . . since there are variables in the denominators on the left side. But if we multiply both sides by x, we can turn this equation into the linear equation

$2 + 1 = 6x$

The bottom line is to make sure you look and simplify your equations carefully before deciding whether or not they can be classified as linear!

WATCH OUT!

◀ Q&DT

SINGLE VARIABLE LINEAR EQUATIONS

Notice that each of the linear equations from the last Pop Quiz contain only one variable. But, you might be thinking, hasn't every equation we've looked at had only a single variable? Yes, that's true . . . although that's all going to change a little later in this chapter. But before we get to that brave new world, let's take a more in-depth look at what are called single variable linear equations. That is, linear equations with only one variable.

POP QUIZ:

ENGLISH TO EQUATION TRANSLATION

It's time to take another shot at translating from English to algebra. But this time instead of an English to expression translation, let's go for an English to equation translation. Here's the sentence in English that I want you to translate into an algebraic equation—in particular, into a single variable linear equation:

The mean of 10 and some number larger than 10 is equal to 1 more than the range of these two numbers.

The first step in doing this translation is to make sure you understand what all the words mean. In this case, I've sprung two new ideas on you that we haven't talked about yet: "mean" and "range."

ALGEBRA DECODER!

• MEAN AND RANGE

I'm going to assume you've learned about the mean and range before . . . probably back when you learned about averages in school. But I'm also going to assume that a quick reminder won't hurt. So, here's a brief summary of their meaning . . .

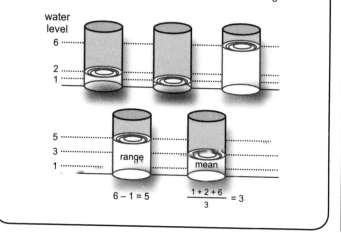

Cont'd

Cont'd from page 166

Mean: Imagine pouring water back and forth between the three glasses on the top above until they all contain equal amounts of water. That new height is called the mean height, and it can be calculated by adding all the heights in the individual glasses and then dividing by the total number of glasses. In this case, that's $(1+2+6) / 3 = 3$. The mean is one type of average, and therefore tells us about the typical height of water in the glasses.

Range: The range tells you the difference in height between the glasses with the most and least amounts of water in them. In this case, the range is equal to $6-1=5$, which tells us something about how much the various heights of water in the glasses are spread out.

And with that explanation, you should be ready to perform the English to equation translation. If you're having trouble, you can always check out the solution at the end of the book. So, once again, here's the sentence:

The mean of 10 and some number larger than 10 is equal to
1 more than the range of these two numbers.

And here's your translated equation:

_____

HOW TO SOLVE SINGLE VARIABLE LINEAR EQUATIONS

If you successfully made it through the "How to Solve an Equation" tutorial in the last chapter, then you're already well on your way to understanding this tutorial too . . . because the general process of solving single variable linear equations is very similar. So let's work through the four-step process from the last chapter to solve the equation we just came up with relating the mean and range of two numbers:

$$\frac{10+x}{2} = x - 10 + 1$$

But the steps here aren't going to be exactly the same as they were for solving equations in general. In particular, watch out for an all-new step at the end of the tutorial.

■ STEP 1: SIMPLIFY EACH SIDE OF THE EQUATION

Let's start simplifying the equation by combining the pieces on the right side to get $x - 10 + 1 = x - 9$. Then let's multiply both sides of the equation by 2 to get

$$10 + x = 2 \bullet (x - 9)$$

How do we simplify the right side now? We can use something called the distributive property. We'll look at this in detail in the next chapter, but here's the quick and dirty explanation. If you have a stick that's c units long (the unit can be whatever you want—meters, miles, or whatever), and if you then divide that stick up into two parts that are each a and b units long, then the length of the stick can also be described as being $a + b$ units long. The picture looks like this:

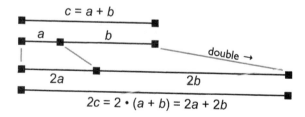

If you then double the length of the stick, you can write the new length as either $2c$ (twice the original length) or as

2 • $(a+b)$, which you can see in the drawing must have a length $2a+2b$. And it works for subtraction too. In other words, if $c=a-b$ instead of $a+b$, then $2c=2 • (a-b)=2a-2b$. All you have to do is substitute x for a and 9 for b into this last equation, and you'll see that the right side of our equation can be simplified like this:

$$2 • (x-9)=2x-2 • 9=2x-18$$

So our fully simplified equation is

$$10+x=2x-18$$

■ STEP 2: MOVE VARIABLE TO ONE SIDE (VIA ADDITION AND SUBTRACTION)

The name of the game now is to get all of the parts of the equation that contain x alone on the same side of the equation. So, let's start by subtracting x from each side of the equation to get all the pieces containing x on the right:

$$10+x-x=2x-18-x$$

Simplifying this gives us

$$10=x-18$$

Now, to get the pieces of the equation with x alone on the right side of the equation, we just need to add 18 to both sides:

$$10+18=x-18+18$$

And once we simplify this, we end up with $28=x$ . . . which, of course, means the exact same thing as $x=28$. That's our answer!

But let's pretend for a minute that we didn't start this step off by isolating the variable on the right side of the equals sign. Let's see what would have happened if we had instead decided to isolate x on the left side of the equation. Let's start back at $10+x=2x-18$, but this time let's subtract $2x$ from both sides:

$$10+x-2x=2x-18-2x$$

which we can simplify to get

$$10-x=-18$$

Now, to get x alone on the left side of the equation, we need to subtract 10 from both sides:

$$10-x-10=-18-10$$

which after simplifying becomes

$$-x=-28$$

We haven't quite solved for x here (we've actually solved for $-x$), but that's as far as we can get in solving for x using addition and subtraction alone. Which means that it's time to move to the third step.

■ STEP 3: ISOLATE THE VARIABLE (VIA MULTIPLICATION AND DIVISION)

In the tutorial on solving equations that we went through in the last chapter, this step could have involved multiplication, division, exponentiation, or taking roots. But when it comes to solving linear equations, you'll never need to use exponentiation or take any roots since the variables in linear problems are never raised to a power other than one. In this case, we can solve for the variable x by multiplying (or by dividing . . . either will work) each side of the equation by –1:

$$-x \bullet -1 = -28 \bullet -1$$

which of course gives us the answer $x=28$. No big shock . . . it's the same answer that we got when we solved the problem the other way around. This goes to show you that as long as you obey the rules of solving equations that we outlined in the last chapter (namely not to throw the equation out of balance), the particular order of steps that you use doesn't matter . . . you'll always get the same correct answer.

■ STEP 4: CHECK YOUR SOLUTION!

Now that we've solved for x (in two different ways, no less!), let's try plugging $x=28$ back into our original equation to make sure that it really works:

LHS $\rightarrow$ $(10+28) / 2 = 38 / 2 = 19$

RHS $\rightarrow$ $28-10+1 = 18+1 = 19$ ✔ (LHS = RHS)

So what does the solution $x=28$ mean? Well, the original equation we came up with says that the mean of 10 and some other number is 1 more than the range of those numbers. And our solution says that the other number must be 28. In other words, the mean of 10 and 28 is 19, and the range of those two numbers is $28-10=18$ . . . which is 1 less than the mean—exactly as our equation says it must be!

■ STEP 5: VISUALIZE YOUR SOLUTION!

Now that we know about number lines, we don't have to be content with simply solving for the value of the variable x . . . we can visualize its place in the wider world of numbers too by showing the solution on the number line.

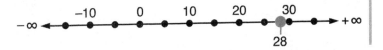

Admittedly there's not a lot to be gleaned from this particular number line, but as we'll see a little later in the chapter, looking at solutions on the number line can be a great way to help you understand the meaning of the solutions to your problem.

To summarize what we've learned in this tutorial, here are the five steps for solving single variable linear equations:

Q&DT ▶

**The 5 Steps of Solving a
Single Variable Linear Equation**

1. Simplify both sides of the equation.
2. Move all parts of the equation that contain the variable you're solving for to the same side.
3. Isolate the variable with multiplication or division.
4. Check your solution!
5. Visualize your solution!

OVERACHIEVER 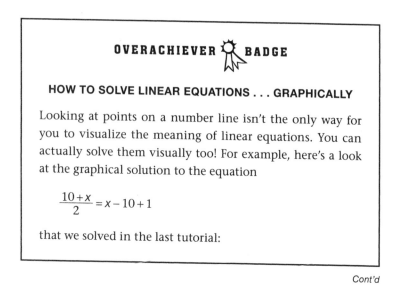 BADGE

HOW TO SOLVE LINEAR EQUATIONS . . . GRAPHICALLY

Looking at points on a number line isn't the only way for you to visualize the meaning of linear equations. You can actually solve them visually too! For example, here's a look at the graphical solution to the equation

$$\frac{10+x}{2} = x - 10 + 1$$

that we solved in the last tutorial:

Cont'd

Cont'd from page 172

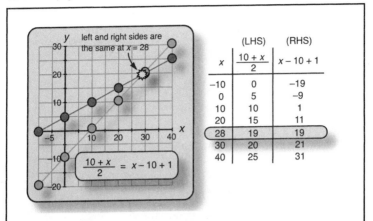

The table on the right contains a list of values for the left and right sides of the equation. The dark gray line on the graph shows the values of the left side of the equation at different x values, and the light gray line shows the values of the right side of the equation. The two lines intersect at the point $x = 28$, which is therefore the point at which the left and right sides of the equation are equal. In other words, $x = 28$ is the solution to the equation. Looking at this graph, it should now be clear why this type of equation is called "linear"—the graphs of both sides are straight lines!

POP QUIZ:

LET'S SOLVE A SINGLE VARIABLE LINEAR EQUATION!

Translate the following sentence into a single variable linear equation:

The mean of 10 and some number less than 10 is equal to 1 more than the range of these two numbers.

_____

Cont'd

Cont'd from page 173

Now use the five-step process for solving single variable linear equations to solve their equation for the unknown value:

Now show your solution on the number line:

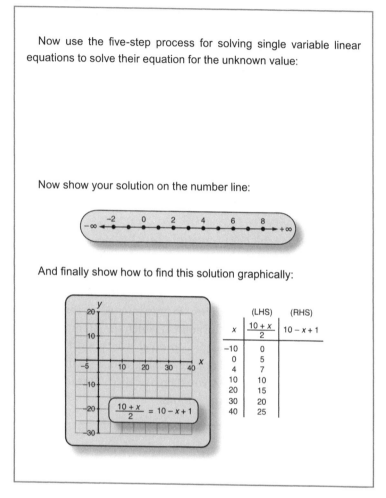

And finally show how to find this solution graphically:

x	$\dfrac{10 + x}{2}$ (LHS)	$10 - x + 1$ (RHS)
−10	0	
0	5	
4	7	
10	10	
20	15	
30	20	
40	25	

$$\frac{10 + x}{2} = 10 - x + 1$$

LINEAR EQUATIONS WITH ABSOLUTE VALUES

Have you noticed anything mildly obnoxious about the two equations we've translated that relate the mean of a pair of numbers to their range? Well, I suppose there could be a number of things you've found annoying, so let me direct your attention to the portions of the equations highlighted here:

*The mean of 10 and some number **larger than** 10 is equal to
1 more than the range of these two numbers.*

$$\frac{10 + x}{2} = (x - 10) + 1$$

these orderings ensure
positive ranges!

$$\frac{10 + x}{2} = (10 - x) + 1$$

*The mean of 10 and some number **less than** 10 is equal to
1 more than the range of these two numbers.*

Notice that the only difference between the two equations is that "$x - 10$" is written in the first and "$10 - x$" in the second. Why is that? Well, in the first equation we're told that the unknown number is larger than 10, and in the second equation we're told that the unknown is smaller. So we had to swap the ordering of the "10" and the "x" in the two equations to ensure that the range is always a positive number (since the range, by definition, must always be positive).

But it's a little cumbersome to always have to be thinking about which order we have to subtract two numbers to make sure the result is positive. And, come to think of it, didn't we learn about a piece of notation earlier that takes care of this exact problem? That's right . . . these equations are just begging to be simplified using absolute values! Instead of these two equations, let's instead write this as the single equation:

$$\boxed{\frac{10 + x}{2} = |x - 10| + 1} = \begin{cases} (x - 10) + 1 & \text{if } x \geq 10 \\ -(x - 10) + 1 & \text{if } x < 10 \end{cases}$$

The absolute value $|x - 10|$ always ensures that the range of the two numbers x and 10 is a positive number. As you can see on the right, by using an absolute value, we've managed to turn two equations into one!

• ALGEBRA TUTORIAL •

HOW TO SOLVE ABSOLUTE VALUE EQUATIONS

Solving linear equations with absolute values is no harder than solving those without . . . except for the fact that there's twice as much work to do. Why is that? To find out, let's walk through the process of solving the absolute value equation we just wrote:

$$\frac{10+x}{2} = |x-10| + 1$$

■ STEP 1: WRITE DOWN THE TWO EQUATIONS

As we've seen, the absolute value on the right side of this equation takes two different forms depending on whether $x > 10$ or $x < 10$. In other words, the two forms depend upon whether or not the contents of the absolute value are positive or negative. When $x \geq 10$, then $x - 10 \geq 0$ and the equation can be written as:

Equation 1: $\dfrac{10+x}{2} = (x-10) + 1$ if $x \geq 10$

When $x < 10$, then $x - 10 < 0$ and the equation can be written as:

Equation 2: $\dfrac{10+x}{2} = -(x-10) + 1$ if $x < 10$

Q&DT ▸

In other words, all you have to do to write the first equation is ignore the absolute value symbols. And all you have to do to write the second equation is to stick a negative sign in front of whatever was inside the absolute value (which ensures that it will have a positive value). Now all that's left for us to do is solve these two equations . . . and then combine the answers to get our final solution.

■ STEP 2: SOLVE THE FIRST EQUATION USING THE FIVE-STEP PROCESS

You've already solved the first equation in the earlier tutorial on How to Solve Single Variable Linear Equations, so we won't repeat the process here. As you'll recall, the solution is $x = 28$.

■ STEP 3: SOLVE THE SECOND EQUATION USING THE FIVE-STEP PROCESS

If you completed the last Pop Quiz, then you've also already solved the second equation (if you haven't, then you might want to do it now). As you'll recall, the solution is $x = 4$.

■ STEP 4: COMBINE BOTH ANSWERS TO GET THE FINAL SOLUTION

So what do those equations mean? Well, they tell you that unlike every other equation we've looked at so far, the equation

$$\frac{10+x}{2} = |x - 10| + 1$$

actually has two solutions! The equation can either be solved by $x = 28$ *or* by $x = 4$. You should plug each of these solutions into the equation to make sure that they both work. Also, let's also go ahead and show this pair of solutions on the number line so that we can see them both together.

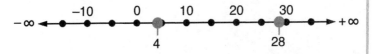

To sum things up, here are the four steps you should follow to solve absolute value equations.

Q&DT ►

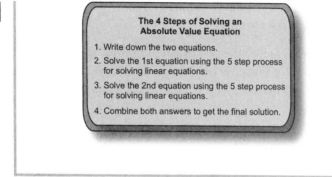

The 4 Steps of Solving an Absolute Value Equation

1. Write down the two equations.
2. Solve the 1st equation using the 5 step process for solving linear equations.
3. Solve the 2nd equation using the 5 step process for solving linear equations.
4. Combine both answers to get the final solution.

OVERACHIEVER BADGE

HOW TO SOLVE ABSOLUTE VALUE EQUATIONS . . . GRAPHICALLY

In the same way that we earlier solved a linear equation graphically, we can also solve absolute value equations graphically. For example, here's a look at the graphical solution to the equation

$$\frac{10+x}{2} = |x-10|+1$$

that we solved in the last tutorial:

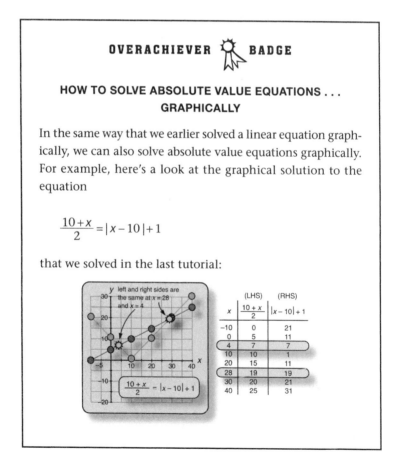

x	(LHS) $\frac{10+x}{2}$	(RHS) $|x-10|+1$
−10	0	21
0	5	11
4	7	7
10	10	1
20	15	11
28	19	19
30	20	21
40	25	31

The table on the facing page contains a list of values for the left and right sides of the equation. Notice that the absolute value on the right side of the equation keeps all of its values positive—and gives its graph on the left a "V" shape. The left side of the equation is shown as the dark gray line, and the right side is shown as the light gray line. The two lines intersect at the points $x=4$ and $x=28$, which are therefore the points at which the left and right sides of the equation are equal. In other words, $x=4$ and $x=28$ are both solutions to the equation. Looking at the graph, it should be clear why we get two solutions for absolute value equations.

 POP QUIZ:
LET'S SOLVE AN ABSOLUTE VALUE EQUATION!

Use the four-step process for solving absolute value equations to solve $| x - 3 | - 1 = 2$ for the unknown value(s) of x:

Show the solution(s) on the number line:

LINEAR INEQUALITIES

It's time for us to put a different spin on the series of problems we've been investigating about the mean and range of pairs of numbers. Until now, we've been looking at an equation like

$$\frac{10+x}{2} = x - 10 + 1$$

which allows us to find the greater-than-10 number that when paired with 10 produces a mean value that's 1 more than the range of the pair. Or we've been looking at an equation like

$$\frac{10+x}{2} = |x - 10| + 1$$

which allows us to find *all* the numbers (whether or not they're greater than 10) that pair up with 10 to satisfy the same conditions about their mean and range.

But what if you don't care about the mean of the pair of numbers being exactly equal to one more than the range. What if instead we ask the question:

*What greater-than-10 numbers can you pair up with 10 so that the mean of the pair is always **less than** 1 more than their range?*

I know this is a mouthful of a sentence, but let's take a shot at translating it into algebra. Here's the answer:

$$\frac{10+x}{2} < x - 10 + 1$$

As you can see, this is very similar to the equation above . . . the only difference is that instead of an equals sign, we now have a less than sign. Which means that what we've just written isn't actually an equation (since there isn't an equals sign), but is instead what's called an inequality.

POP QUIZ: EQUALITY VS. INEQUALITY

With the "translation" from the equation on the left side of the following picture to the inequality on the right in mind, translate the sentences below into inequalities.

Cont'd

Cont'd from page 180

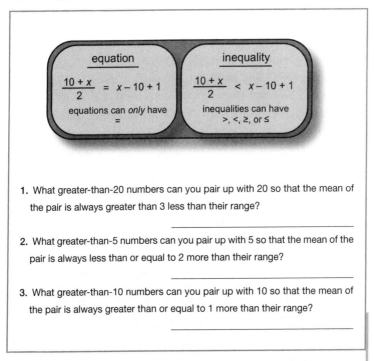

1. What greater-than-20 numbers can you pair up with 20 so that the mean of the pair is always greater than 3 less than their range?

   _____

2. What greater-than-5 numbers can you pair up with 5 so that the mean of the pair is always less than or equal to 2 more than their range?

   _____

3. What greater-than-10 numbers can you pair up with 10 so that the mean of the pair is always greater than or equal to 1 more than their range?

   _____

• **ALGEBRA TUTORIAL** •

HOW TO SOLVE LINEAR INEQUALITIES

So how do we solve an inequality like the one we just before the last Pop Quiz? Well, as we've discovered, inequalities look an awful lot like equations . . . which means that the process of solving them is pretty similar to the process of solving equations. To see what I mean, let's take a few minutes and walk through the process of solving the linear inequality

$$\frac{10+x}{2} < x - 10 + 1$$

▪ STEP 1: SIMPLIFY BOTH SIDES OF THE INEQUALITY

Just as with solving equations, the first step in solving inequalities is to simplify the expressions on both sides of the inequality sign (whatever that may be . . . in this case it's "<"). So, let's first simplify the right side to find that $x - 10 + 1 = x - 9$, and then let's multiply both sides by 2. Using the distributive property that we talked about in the tutorial "How to Solve Single Variable Linear Equations," we find that $2 \cdot (x - 9) = 2x - 18$, and our inequality therefore simplifies to

$$10 + x < 2x - 18$$

This should look familiar! The only difference so far between solving this inequality and the similar equation earlier is that we've swapped the equals sign and thrown in a less than sign.

▪ STEP 2: MOVE VARIABLE TO ONE SIDE (VIA ADDITION AND SUBTRACTION)

The next step of solving inequalities should also look familiar from solving equations . . . in fact, it's the exact same step! Our goal is simply to get all of the parts of the inequality that contain x alone on one side of the inequality sign or the other. If we move all parts containing x to the right by subtracting x from both sides, we get

$$10 + x - x < 2x - 18 - x$$

which after simplifying becomes

$$10 < x - 18$$

Now all we have to do is add 18 to both sides of the inequality to get the variable x alone on the right side:

$$28 < x$$

And this is our solution . . . which we can turn around and write as $x > 28$. But before we discuss what this solution means, let's see what would have happened if we had decided to move everything containing an x to the left side of the inequality instead. To do that, we need to subtract $2x$ and 10 from both sides of $10 + x < 2x - 18$, like this:

$$10 + x - 2x - 10 < 2x - 18 - 2x - 10$$

After simplifying this, we get

$$-x < -28$$

Which means that to finish solving the inequality this way, we need to move on to step 3.

■ STEP 3: ISOLATE THE VARIABLE . . . AND MAYBE SWITCH THE INEQUALITY (VIA MULTIPLICATION AND DIVISION)

Why did I make you slog through solving this inequality two ways? I did it because I wanted to get to this third step . . . because this is where solving inequalities is a little different than solving equations. How? Well, this is the step in the equation and inequality solving process where we multiply or divide by some number to isolate the variable. And sometimes that number we multiply or divide by is a negative number. While that's a perfectly fine thing to do when you're solving equations, something funny happens with inequalities. When you multiply or divide an inequality by a negative number, you fundamentally change the meaning of the inequality—definitely something to watch out for . . .

Multiplying and Dividing Inequalities by Negative Numbers

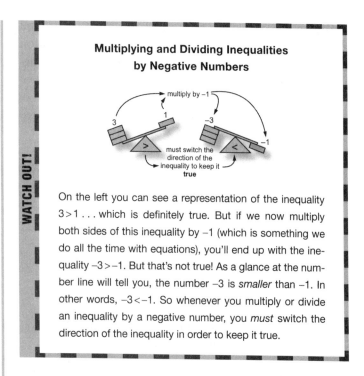

On the left you can see a representation of the inequality $3 > 1$ . . . which is definitely true. But if we now multiply both sides of this inequality by -1 (which is something we do all the time with equations), you'll end up with the inequality $-3 > -1$. But that's not true! As a glance at the number line will tell you, the number -3 is *smaller* than -1. In other words, $-3 < -1$. So whenever you multiply or divide an inequality by a negative number, you *must* switch the direction of the inequality in order to keep it true.

What does this mean for our problem? Well, at this point our problem looks like $-x < -28$. In order to isolate x on the left side, we need to multiply both sides of the inequality by -1. And when we do that, we need to remember to switch the less than symbol into a greater than symbol, like this:

$-x \bullet -1 > -28 \bullet -1$

which simplifies to give us

$x > 28$

the exact same solution we found when we solved the problem the other way around!

■ **STEP 4: CHECK YOUR SOLUTION!**

Just because this is an inequality instead of an equation doesn't mean we can skip the step of checking our solution. It's *always* a good idea to plug your solution in and make sure it's true. In this case, our solution says that all numbers greater than 28 should work. So let's try plugging $x=30$ into our original inequality:

LHS $\rightarrow$ $(10+30) / 2 = 40 / 2 = 20$
RHS $\rightarrow$ $30 - 10 + 1 = 20 + 1 = 21$

Which says that $20 < 21$. That's true . . . our solution works! Try plugging in a few more numbers both greater and less than 30 to make sure the solutions continue to hold up. What happens when we set $x=28$? Well, in that case both the LHS and RHS equal 19. But since 19 is not less than 19, the solution $x=28$ does *not* work . . . which is consistent with our solution that $x > 28$.

■ **STEP 5: VISUALIZE YOUR SOLUTION!**

When solving linear equations, we always ended up with one solution . . . or two in the case of absolute value equations. But with inequalities, we don't just get one or even two solutions, we get a whole continuum of them. In other words, every number that's greater than 28 is a solution to this problem. It doesn't matter if the number is an integer, a rational number, or an irrational number—if it's greater than 28, it's a solution. But how can we visualize these solutions? Well, with one or two solutions we drew points on the number line. With the continuum of solutions that we get with inequalities, we do this:

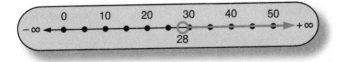

Start by drawing a circle on the number line at the endpoint of your solution—in our case that's at 28. If the solution includes the endpoint—that is, if $x = 28$ is a valid solution to the inequality—then fill in the circle. If the endpoint is not included (as in our problem since the solution is $x > 28$ and not $x \geq 28$), don't fill in the circle. Finally, draw a solid line and an arrow on top of the number line indicating where all the rest of the solutions are.

To sum up, whenever you need to solve an inequality, just remember to follow this five-step plan:

Q&DT ▶

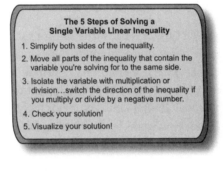

**The 5 Steps of Solving a
Single Variable Linear Inequality**

1. Simplify both sides of the inequality.
2. Move all parts of the inequality that contain the variable you're solving for to the same side.
3. Isolate the variable with multiplication or division…switch the direction of the inequality if you multiply or divide by a negative number.
4. Check your solution!
5. Visualize your solution!

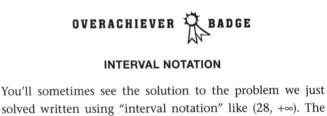

OVERACHIEVER BADGE

INTERVAL NOTATION

You'll sometimes see the solution to the problem we just solved written using "interval notation" like $(28, +\infty)$. The number on the left tells you the lower limit for the solution, and the number on the right tells you the upper limit. If the lower or upper limits are included in the solution, then these endpoints are called "closed" and you use "[" or "]" symbols at the ends of the interval. If they are not, then these endpoints

Cont'd

Cont'd from page 186

are called "open" and you use "(" or ")" symbols at the ends of the interval. You can indicate that the solution extends indefinitely in one direction or the other using the symbols for positive and negative infinity: $+\infty$ or $-\infty$ (these endpoints are always written as open intervals). Here are a few examples:

- $x < 28$ corresponds to $(-\infty, 28)$
- $x \leq 28$ corresponds to $(-\infty, 28]$
- $x > 0$ and $x \leq 5$, also known as $0 < x \leq 5$ corresponds to $(0, 5]$
- $x \geq -10$ corresponds to $[-10, +\infty)$

POP QUIZ:
INEQUALITY TO INTERVAL NOTATION
TRANSLATION

Convert the following inequalities into interval notation:

1. $-1 < x \leq 1$ .... _____
2. $4 \leq x < 14$ .... _____
3. $10 < x$ .... _____
4. $x \leq -9$ .... _____

| M | A | T | H | | B | R | A | I | N | | G | A | M | E | |

PICTURING ABSOLUTE VALUE EQUATIONS
AND INEQUALITIES

It's now time for us to put together everything we've learned so far about absolute values and inequalities to tackle a math brain game. In particular, the goal of the game I'm thinking about is to figure out which of these

$$|x+1| = 2 \qquad |x| < 3$$

$$|x| = 1$$

$$|x-1| = 2 \qquad |x| \geq 3$$

produce each of these

As we've discussed, solid gray dots and lines indicate points and regions of the number line that are solutions to an equation or inequality. Unfilled gray dots indicate points that are not solutions to an inequality. Take a minute to think about these equations and inequalities and see if you can figure out which number line goes with each of them. And then check out the answers and explanations.

MATH BRAIN GAME SOLUTIONS

Without further ado, here are the solutions to our brain game. Each equation or inequality is written above the number line showing its solutions.

PUZZLE #1

$$|x| = 1$$

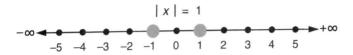

As indicated by the solid gray dots, the numbers –1 and 1 are both solutions to the equation $|x| = 1$. How do we know? Well, let's check and see: $|-1| = 1$ and $|1| = 1$. Yes, they both work! Of course, we could also go through our formal solution method to figure this out. The first step is to write down the two equations represented by the absolute value equation. In this case, when $x \geq 0$, the equation is just $x = 1$. There's the first solution! And when $x \leq 0$, we get the other equation: $-x = 1$. If we multiply both sides by –1, we get the second solution: $x = -1$. Remember that the absolute value of a number simply tells us how far away it is from 0. Similarly, in this case the absolute value of $|x|$ represents how far away the value of x is from 0. So the problem $|x| = 1$ is really just asking us to find all the numbers that are 1 away from 0. And, of course, those numbers are –1 and 1.

PUZZLE #2

$$|x - 1| = 2$$

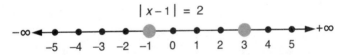

Again, we can use the skills we've learned to solve the absolute value equation $|x - 1| = 2$. When $x - 1 > 0$, this equation is equal to $x - 1 = 2$ . . . which has the solution $x = 3$. And when $x - 1 < 0$, the absolute value equation is equal to $-(x - 1) = 2$. Multiplying both sides by –1 and then adding 1 to each side gives us the second solution $x = -1$. But what do these solutions mean? Well, if $|x|$ tells us how far away the value of the variable x is from 0, then $|x - 1|$ tells us how far away the value of the variable x is from 1. Do you see why? If we plug in $x = 1$, we get 0; and if we plug in either $x = 0$ or 2, we get 1, and so on— the answer is always the distance that x is away from 1. So, since $|x - 1|$ tells us how far away x is from 1, $|x - 1| = 2$ is simply asking us to find all the locations on the number line that are 2 away from 1. Not surprisingly, those locations are just $1-2=-1$ and $1+2=3$.

PUZZLE #3

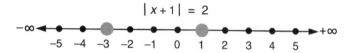

$$|x+1| = 2$$

First of all, the equation $|x+1|=2$ is very similar to the previous equation. The only difference is that we now have $|x+1|$ instead of $|x-1|$. What does that mean? Well, it means that we can solve this equation to find its solutions, or we can take a shortcut based upon what we learned about the meaning of the solutions in the last problem. Specifically, whereas $|x-1|$ tells us how far away the variable x is from 1, $|x+1|$ tells us how far away it is from –1. Go ahead and plug in a few numbers to make sure this is true. So, $|x+1|=2$ is simply asking you to find all the locations on the number line that are 2 away from –1. Those locations are just $-1-2=-3$ and $-1+2=1$.

PUZZLE #4

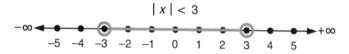

$$|x| < 3$$

Let's go ahead and apply our absolute value inequality solving skills to find the region of the number line that makes $|x|<3$ true. When $x\geq0$, this inequality is just $x<3$. And when $x<0$, this inequality is just $-x<3$. Now, when we multiply both sides by –1, we must remember to switch the less than symbol to a greater than symbol. Which gives us $x>-3$. When we put these two solutions together, we see that the inequality is true whenever x is both greater than –3 (but not equal to –3) and less than 3 (but not equal to it). We can write these two conditions more succinctly as $-3<x<3$, or in interval notation as $(-3, 3)$.

PUZZLE #5

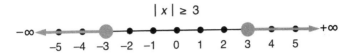

Let's figure out the solution to this inequality by thinking about what it is telling us. As we've discussed, $|x|$ represents the distance from x to 0. Which means that $|x| \geq 3$ is simply asking you to find all the regions of the number line that are at least 3 away from 0. Of course, these are just the two regions $x \leq -3$ and $x \geq 3$. Or, in interval notation, the solution is $(-\infty, -3]$ and $[3, +\infty)$.

OVERACHIEVER ⚝ BADGE

LINEAR EQUATIONS WITH TWO VARIABLES

Lest you think that all linear equations in algebra have only one variable, let's take a look at a problem that will quickly turn this idea on its head. It's a problem that you're already familiar with . . . but with one key difference. Let's say that you're interested in finding a pair of numbers whose mean is one more than their range. But instead of assuming that the value of one of the two numbers is 10 (as we've done throughout the chapter), let's leave both their values completely unknown and instead call them x and y. Which means that the relationship we're after can be expressed by the equation

$$\frac{x+y}{2} = y - x + 1$$

Cont'd

Cont'd from page 191

Just to keep things simple here, I've written this equation assuming that $y > x$ . . . since the range, $y - x$, has to be a positive number (you could have instead used absolute values as we've done before). To solve this equation, start by multiplying both sides by 2 to get

$x + y = 2y - 2x + 2$

Now let's get y all alone on the right side of the equation by subtracting y and 2 from both sides of the equation, and by adding $2x$ to both sides of the equation. The result is:

$3x - 2 = y$

Let's turn this around and write it as

$y = 3x - 2$

What does this mean? Well, the real answer to that question is going to come in a few chapters when we get very well acquainted with equations like this. But let's get a flavor for the answer now by making a graph with x on the horizontal axis and $y = 3x - 2$ on the vertical axis:

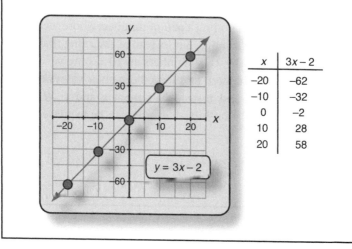

x	$3x - 2$
−20	−62
−10	−32
0	−2
10	28
20	58

Cont'd

Cont'd from page 192

This is the graph of the line $y = 3x - 2$. The cool thing about this line is that every single point on it is a solution to our problem. In other words, each point on the line gives us an ordered pair of numbers that is a solution to our original problem of finding two numbers whose mean is one greater than their range. For example, at $x = 15$ we get $y = 3 \cdot 15 - 2 = 43$. If you check, you'll see that the mean of 15 and 43 is indeed equal to one more than their range. Which means that $x = 15$ and $y = 43$ are one of an infinite number of solutions to our problem!

WRAP-UP

So that's the story of numbers, the line they live on, and the lines we find in algebra resulting from linear equations and inequalities. Now that we've covered all of these critical components of any good algebraic repertoire, it's time for a poem.

Numbers, numbers, everywhere
They're in the trees, they're in the air
But no matter where you look, you'll find
Those numbers are all in your mind

What do you think? Pretty good, right? Yeah, I'm so proud of it that I'm thinking of writing the next chapter in verse . . . with a quill and ink no less. On second thought, perhaps I'd better stick with my nicely typed-up prose since we're going to be talking about the important ideas that will help us smoothly transition from the friendly confines of the linear world we've been playing in to the great big nonlinear world at large.

· FINAL EXAM ·

• NUMBERS, THE NUMBER LINE, AND ABSOLUTE VALUES

1. Using the following number line

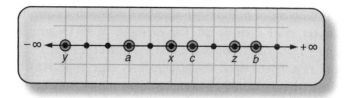

write an equation that says that the distance between *x* and *y* is equal to the difference of two absolute values.

• SOLVING LINEAR EQUATIONS

2. Translate the following sentence into a single variable linear equation:

The mean of 10 and some number greater than 10 is equal to 1 less than the range of these two numbers.

Now use the five-step process for solving single variable linear equations that we talked about earlier to solve this equation for the unknown value:

Finally, show the solution to this linear equation on the number line:

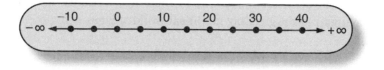

- **SOLVING ABSOLUTE VALUE EQUATIONS**

3. Draw a graph that shows how to visually find the solution(s) of the absolute value equation

$$|x-3|-1=2$$

that we looked at in the Let's Solve an Absolute Value Equation! Pop Quiz:

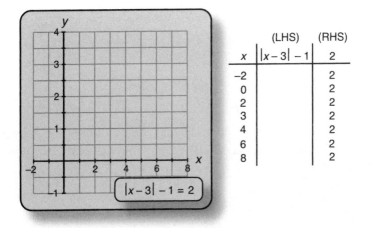

- **SOLVING INEQUALITIES**

4. Use the five-step process for solving linear inequalities to solve the problem

$$\frac{10+x}{2} \geq |x-10|+1$$

Oh no! That's an inequality with absolute values! Yes, I'm throwing a double whammy at you. But I'm confident that you can handle it. And if you get a little stuck, just check out the solution at the end of the book. So how should you start? Well, if you think about it, you'll see that we actually have two inequalities to solve here since the absolute value has two cases: one where the stuff inside it is positive, and the other where it's negative. Which means you need to solve these two inequalities:

Inequality 1: $\dfrac{10+x}{2} \geq x - 10 + 1$ if $x \geq 10$

Inequality 2: $\dfrac{10+x}{2} \geq -(x - 10) + 1$ if $x < 10$

Okay, I've said enough . . . I'm going to leave you alone now to solve the problem! Solve the first inequality here:

And the second here:

Then combine your two solutions and write the final solution in interval notation.

_____

Finally, show your complete solution on the number line:

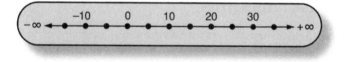

What does this all mean? Well, the values of x in the solution are *all* the numbers (both larger and smaller than 10) that when paired with 10 have a mean that's at least as large as 1 more than their range! This goes to show you that even though algebra equations can start to look (and sound) very complicated, they always have some real underlying meaning at their core.

CHAPTER 4

Arithmetic 2.0: Math with Variables, Exponents, and Roots

Can you do Division? Divide a loaf by a knife—what's the answer to that?

—Lewis Carroll, *Through the Looking Glass*

Congratulations! You've made it to the halfway point of the book. If you take a look back at everything we've covered so far, you should be feeling pretty good about yourself. Hopefully you've acquired a new outlook and perspective on math, learned the basics of algebra, and now have a solid understanding of the roots of the subject: numbers, number lines, linear equations, linear inequalities, and more.

Now, in line with the theme of this part of the book, it's time for us to once again press pause on our story and rewind back to the roots of our subject to take a look at good old arithmetic— what I'm calling arithmetic 1.0. Remember our discussion of not burning down the kitchen in chapter 3? This section is here for that same reason—I'm going to explain many math 1.0 concepts (like addition, subtraction, multiplication, and division) to you in a new way so that you'll have a deeper understanding that will help you when it comes to algebra. So again: Even though this material might seem like one big old review, don't burn down the kitchen by skipping ahead! Your future self will thank you.

Then, with new math knowledge attained, we'll once again start up our forward march toward understanding how the relatively simple ideas of arithmetic work their way into the world of algebra in the form of what I'm calling "arithmetic 2.0" . . . which means arithmetic with variables, exponents, and roots.

ARITHMETIC 1.0: PREPARING FOR ALGEBRA

At this point in your life, you're something of an arithmetic expert. You've been doing things like adding, subtracting, multiplying, and dividing numbers for so long that you can pretty much do it in your sleep. But this stuff, what I'm calling arithmetic version 1.0, is really only the tip of the math iceberg. As we'll see later in this chapter, understanding all the basics of math is incredibly important for understanding algebra, or arithmetic 2.0. So, to make sure we're all prepared for that, let's take a few minutes to look back at arithmetic 1.0.

ADDING POSITIVE INTEGERS

In many ways mathematics begins with addition. As you know, addition is just the process of combining things into bigger groups of things. Those things could be piles of rocks, dollars, pumpkins, or whatever else you want to put together into bigger piles, like this:

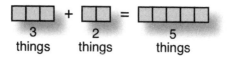

Well, duh! Yes, indeed—duh. But that's addition!

SUBTRACTING POSITIVE INTEGERS

Okay, how about subtraction? Well, it's just the process of taking one big pile and taking away some portion of it to make a smaller pile. In other words, it's the exact opposite of addition—one adds things to piles and the other removes things from them, like this:

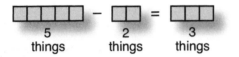

Yes, this is another one of those "duh" moments. But, once again, that's subtraction!

ALGEBRA DECODER!

• **INVERSE PROCESS**

In math, something that undoes what something else does is called an **inverse process.** Which means that subtraction is the inverse process of addition. One way to think of this is that we can add and subtract the same number from another number without changing the total. In other words, $1 + 10 - 10$ is still equal to 1! We'll see several other examples of inverse processes in this chapter.

MULTIPLYING POSITIVE INTEGERS

There are a few ways to think about multiplying numbers. One way is to think of multiplication as a process that scales one number until it is some other number of times its original size. For example, you can think of the problem $3 \cdot 5$ as saying that the number 5 is scaled up until it is three times bigger than it was to begin with. In other words, take five blocks and stretch them out until they are three times their original size, like this:

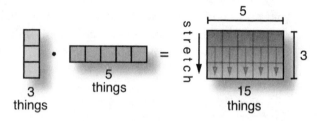

What do you end up with? Not surprisingly, you get a 3-by-5 rectangular group of blocks. How many blocks is that in all? Of course it's just 3 • 5 = 15. Which means that we can always think of the product of two numbers as the area of a rectangle.

ALGEBRA DECODER!

• **AREA**

The concept of **area** is used all the time in everyday life. For example, let's say you need to figure out how much money new carpeting for your bedroom will cost. When you go to the store, you see that it's sold by the square foot. So to find the total cost for your room, you need to multiply the price per square foot by the total number of square feet in your bedroom.

Imagine taking a bunch of squares of carpet that are 1 foot long on each side and laying them out side by side so they completely cover the floor. The total number of squares you need to do this is exactly equal to the area in square feet. If your room happens to be rectangular, you can take a shortcut to finding the area by multiplying the length of the room in feet by its width . . . just like we did when multiplying 3 • 5 = 15.

DIVIDING POSITIVE INTEGERS

Remember talking about the fact that subtraction is the inverse process of addition? Well, it turns out that multiplication has an inverse process too—it's called division (which I'm sure you've heard of). Just as we can picture 3 • 5 as a process that stretches the number 5 until it's three times bigger than its original size, we can picture the problem 15 / 3 as a process that compresses the number 15 until it's three times smaller than its original size, like this:

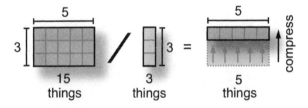

As you can see, we can think of 15 / 3 as a process that takes a rectangular group of 3 rows of 5 blocks and compresses it down until it's three times smaller than its original size. In other words, the process ends up producing a single row of 5 blocks. Which means that 15 / 3 = 5.

DOING ARITHMETIC WITH "REAL" NUMBERS

You may have noticed that we've only been talking about doing arithmetic with positive integers so far. Why? Well, there's something I haven't mentioned yet about all these nice pictures we've been using to think about good old-fashioned arithmetic. And that's that everything gets a lot more complicated once we start talking about negative numbers and fractions. Okay, maybe things don't get a *lot* more complicated, but they do get a bit more complex. Let's take a look and see what I mean.

ADDING AND SUBTRACTING FRACTIONS

Learning to work with fractions is really a topic for a pre-algebra book, so we're not going to go through every last detail about working with fractions. But we are going to talk a bit about what arithmetic with fractions really means . . . and how to do the most important things. Let's start by thinking about how we add and subtract fractions. In particular, let's look at $\frac{1}{2} + \frac{1}{3}$ and $\frac{1}{2} - \frac{1}{3}$:

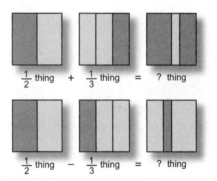

As you can see, you can begin to get a sense of the answers to these problems by visualizing adding to and cutting out pieces from a whole. But to get numerical answers, you need to redraw these fractions using what's called a common denominator. In other words, you need to rewrite the fractions so that they're expressed as some number of the same somethings, like this:

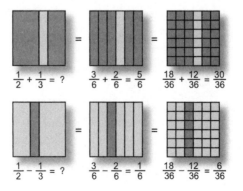

This shows that we can rewrite our fractions as some number of sixths (in the middle) or some number of thirty-sixths (on the right), and then it becomes much easier to do the arithmetic—we just have to add up the boxes. In these problems, 6 and 36 are two common denominators for the fractions $\frac{1}{2}$ and $\frac{1}{3}$. Notice that there isn't just one common denominator, there are actually an infinite number of them—and each of them can be used to give you an equivalent version of the fraction. The easiest way to find a common denominator is to multiply all the denominators of the fractions that you're adding or subtracting.

◀ Q&DT

POP QUIZ:

LET'S ADD AND SUBTRACT SOME FRACTIONS!

Just to make sure you've got this down, try dividing up the following boxes to help you add and subtract these fractions.

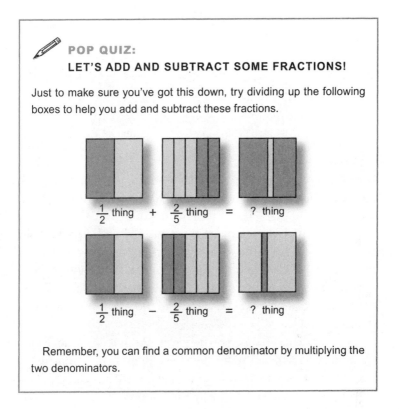

$\frac{1}{2}$ thing   + $\frac{2}{5}$ thing   =   ? thing

$\frac{1}{2}$ thing   − $\frac{2}{5}$ thing   =   ? thing

Remember, you can find a common denominator by multiplying the two denominators.

So that's really all that addition and subtraction with fractions boils down to. Just as with addition and subtraction with positive integers, arithmetic with fractions is essentially nothing more than stacking and removing rocks from a pile . . . except that with fractions the rocks can be broken into pieces!

MULTIPLYING FRACTIONS

We saw earlier that we can think of multiplying and dividing positive integers as ways of either stretching or compressing the magnitude, or value, of a number. In other words, if we multiply 5 by 2 we stretch the magnitude of 5 until it's twice as big . . . giving us 10. But what about fractions? Well, for the most part things work the exact same way. Well, sort of.

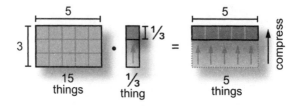

As you can see, this drawing shows the problem $15 \cdot \frac{1}{3} = 5$. And the big news here is that multiplying by $\frac{1}{3}$ does indeed scale 15 . . . to a smaller number. In other words, it actually compresses it! If you look at what we did earlier, you'll see that this is exactly the same thing that happened when we divided 15 by 3. Which, of course, means that dividing by 3 is the same as multiplying by $\frac{1}{3}$. Well that sure does explain why we're using that slashed line notation for division, doesn't it?

So that's what happens when you multiply a number by a fraction with a magnitude that's less than 1, but what if the magnitude of the fraction is greater than 1? For example, how about multiplying something by the fraction $\frac{7}{2}$ . . . also known as $3\frac{1}{2}$?

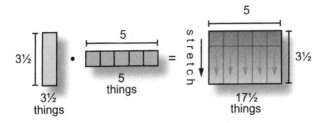

Yes, you get exactly what you would expect. All we've done is stretch the magnitude of the number 5 to be three and a half times bigger than its original size . . . which means that the area of the rectangle on the right must be $3\frac{1}{2} \cdot 5 = \frac{7}{2} \cdot 5 = 35/2 = 17\frac{1}{2}$. If you think about it, you'll see that we can write an algebraic formula for this that will work for any fraction (which we can represent as the ratio of variables $\frac{a}{b}$) and number (which we can represent with the variable c) combination:

$$\frac{a}{b} \cdot c = \frac{ac}{b}$$

OVERACHIEVER 🏅 BADGE

"REPEATED ADDITION"

People frequently think about multiplication as a sort of shortcut for adding a number to itself some other number of times. For example, you can think of the expression $3 \cdot 5$ as $5 + 5 + 5 = 15$:

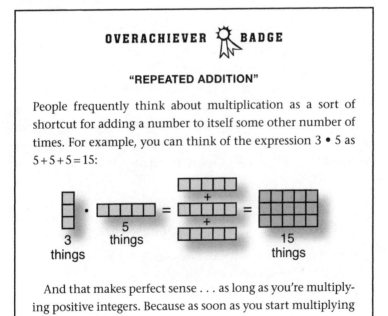

And that makes perfect sense . . . as long as you're multiplying positive integers. Because as soon as you start multiplying

---

Cont'd

Cont'd from page 205

---

fractions, things get a little strange. For example, while it makes perfect sense to add 5 to itself three times, it's definitely a little weird to think about adding 5 to itself three and a half times! But thinking about multiplication as the area of a rectangle that you get by stretching the magnitude of a number always makes sense.

So we've now multiplied integers by fractions that have magnitudes both greater than and less than 1. How about multiplying two fractions? Well, nothing really changes . . . all you're doing is stretching or compressing the magnitude of one fraction by some other fractional amount. Just like we did when multiplying a fraction by a whole number, we can write an algebraic formula that expresses how to multiply any fraction $\frac{a}{b}$ times another fraction $\frac{c}{d}$:

$$\frac{a}{b} \cdot \frac{c}{d} = \frac{ac}{bd}$$

As you can see, a and c represent the numerators of the two fractions, and b and d represent their denominators. The only thing that has changed between this formula and the previous one for multiplying a fraction by a whole number is that we have another denominator to worry about.

POP QUIZ: LET'S MULTIPLY "*n*" FRACTIONS!

That's a nice rule for multiplying two fractions, but why stop now? Instead of two, can you write down a rule for multiplying *n* fractions? What do I mean by *n* here? I mean that you should be able to use this rule to multiply *any* number of fractions . . . 2, 3, 5, 100, a million, it doesn't matter. Check out the solution at the end of the book if you get stuck.

DIVIDING FRACTIONS

When it comes to dividing fractions, most people fall back on the slogan, "Invert and multiply." But what are we really doing when we do that? Actually, we've already discovered the answer to that question! We found earlier that multiplying by a fraction like $\frac{1}{3}$ is the same as dividing by the number 3. But since the number 3 can also be written as the fraction $\frac{3}{1}$, it's clear that dividing by 3 and multiplying by $\frac{1}{3}$ (which is just the inverted version—aka the reciprocal—of $\frac{3}{1}$) is just "invert and multiply":

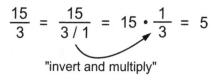

$$\frac{15}{3} = \frac{15}{3/1} = 15 \cdot \frac{1}{3} = 5$$

"invert and multiply"

So, when it comes to dividing fractions, invert and multiply is a good rule to live by. And, similar to what we did when multiplying fractions, we can write a formula that tells us how to use invert and multiply to divide any pair of fractions:

◄ Q&DT

$$\frac{a}{b} \Big/ \frac{c}{d} = \frac{a}{b} \cdot \frac{d}{c} = \frac{ad}{bc}$$

✏️ **POP QUIZ:**
FRACTIONS, ARITHMETIC, AND ALGEBRA

This might surprise you, but you don't really need to remember any of the rules we've learned for adding, subtracting, multiplying, and dividing fractions. Why? Because you can solve all of these problems using algebra. Take a shot at solving these problems for the variable x using the techniques we've developed for solving equations over the past few chapters, and you'll see what I mean.

Cont'd

Cont'd from page 207

1. $\frac{1}{2} + \frac{1}{3} = x$

2. $\frac{1}{2} - \frac{1}{3} = x$

3. $\frac{1}{2} \cdot \frac{1}{3} = x$

4. $\frac{1}{2} / \frac{1}{3} = x$

Of course, the real purpose of our looking at the "rules" of doing arithmetic with fractions wasn't to give you a bunch of formulas to memorize. It was to help you understand what it all really means so that you don't need those formulas in the first place . . . because you can figure them out for yourself!

MULTIPLYING AND DIVIDING NEGATIVE NUMBERS

Okay, we're now comfortable with doing arithmetic with integers and fractions . . . but only positive ones. We haven't talked about what to do with the other half of all those numbers—the negative half. As with fractions, this is really a pre-algebra topic, so we're not going to talk about every situation you'll ever need to worry about. Instead, we're going to focus on understanding what arithmetic with negative numbers means. Up first, let's think about what happens when you multiply a positive number by –1.

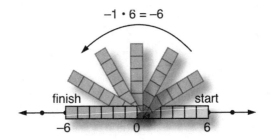

$$-1 \cdot 6 = -6$$

finish   start

−6   0   6

As you can see, all that happens is that the once positive number 6 gets flipped across the origin of the number line to become −6. The size of the number doesn't change since the magnitude (in other words the absolute value) of the number we multiplied by is $|-1| = 1$ . . . which means that no stretching was involved. If we had instead multiplied by something like –3, not only would we have flipped the number across the origin, but we would have stretched it so that its absolute value became three times bigger too.

So multiplying a number by –1 produces the mirror image of that number across the origin. What happens if we now multiply by –1 again? In other words, what happens if we have $-1 \bullet -1 \bullet 6$?

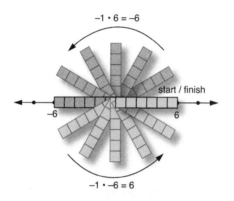

$-1 \bullet 6 = -6$

start / finish

–6   6

$-1 \bullet -6 = 6$

As you can see, this second multiplication by –1 just flips –6 through the origin and puts it right back where it started from. In other words, $-1 \bullet -1 \bullet 6 = 6$ . . . which means that $-1 \bullet -1 = 1$. And now you know exactly where the saying, "A negative times a negative is a positive," comes from.

But wait, there's more! We can take these observations about the effects of multiplying negative numbers and make up a rule to describe them:

- Multiplying an **odd** number of negative numbers gives a **negative** number.
- Multiplying an **even** number of negative numbers gives a **positive** number.

For example, $-1 \bullet -2 \bullet -3$ and $-5 \bullet 10$ both have an odd number of negative numbers (three in the first case and one in the second), so without doing any arithmetic, we already can say that the result must be a negative number. On the other hand, the problem $-1 \bullet -2$ has an even number of negative numbers, so we can immediately say that result must be positive.

What about division? Well, there's not a lot to say because everything works exactly the same for division as it does for multiplication. But just for fun, we can use a little bit of algebra to show that this is true. Specifically, let's solve the equation $x = 5 \, / -1$ to show that dividing a positive number by a negative number results in a negative number. First, let's multiply both sides of the equation by -1 to get $-x = 5$. Then, let's multiply both sides by -1 again to get $x = -5$. And there you have it! So we can write down a rule for division that's similar to the one we wrote for multiplication:

- Dividing an **odd** number of negative numbers gives a **negative** number.
- Dividing an **even** number of negative numbers gives a **positive** number.

In fact, we can do one better than this. Because any combination of multiplying and dividing by a negative number will actually work. In other words, we can say

- Multiplying or dividing an **odd** number of negative numbers gives a **negative** number.
- Multiplying or dividing an **even** number of negative numbers gives a **positive** number.

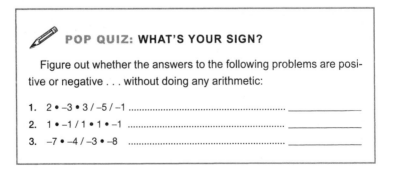

POP QUIZ: WHAT'S YOUR SIGN?

Figure out whether the answers to the following problems are positive or negative . . . without doing any arithmetic:

1. $2 \bullet -3 \bullet 3 \, / -5 \, / -1$ .... _____

2. $1 \bullet -1 \, / \, 1 \bullet 1 \bullet -1$ .... _____

3. $-7 \bullet -4 \, / -3 \bullet -8$ .... _____

ADDING NEGATIVE NUMBERS

The final mini-lesson of this class on doing math with "real" numbers is how to think about adding and subtracting negative numbers. You probably know how to do these types of problems, but it never hurts to be reminded that you can always fall back on the number line to help you figure out the answer. For example, you can see here the various ways that two numbers can be added.

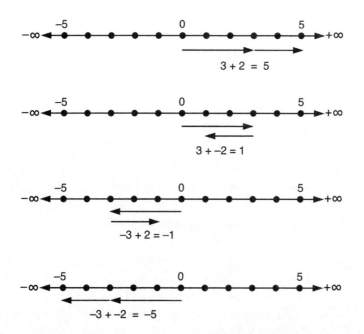

As you can see, a positive number is represented by a positive facing arrow with a length equal to the magnitude (absolute value) of the number, and a negative number is represented by a similar arrow but this time facing in the negative direction. To add positive and negative numbers, all you have to do is lay the arrows tail to head along the number line with the tail of the first arrow starting at zero. Wherever the final head ends up on the number line is the solution to the problem.

SUBTRACTING NEGATIVE NUMBERS

You can think about subtracting negative numbers using a similar setup. But what does it mean to subtract an arrow? Well, perhaps not surprisingly, to subtract an arrow (it can be an arrow representing either a positive or a negative number), you simply flip the arrow around and point it in the opposite direction.

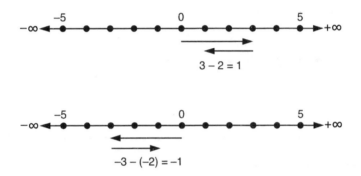

Notice that the arrows in $3-2=1$ look exactly the same as the arrows in $3+(-2)=1$ earlier. And the same thing can be said about the arrows in $-3-(-2)=-1$ and $-3+2=-1$. What's going on? Well, this tells us two very important things:

- Subtracting a positive number from another number has the exact same effect as adding the negative of the number you're subtracting. In other words, $3-2=3+(-2)$.
- Subtracting a negative number from another number has the exact same effect as adding the absolute value of the number you're subtracting. In other words, $-3-(-2)=-3+|-2|=-3+2$.

And the arrows make it perfectly clear why these are true!

POSITIVE AND NEGATIVE OUTCOMES

That just about wraps up our look at how positive and negative numbers work with the big four math operations. But before we completely close the door on this, take a look at the following table that summarizes when a result is positive or negative:

$|x| > |y|$	$x > 0$ $y > 0$	$x > 0$ $y < 0$	$x < 0$ $y > 0$	$x < 0$ $y < 0$
$x + y$	positive	positive	negative	negative
$x - y$	positive	positive	negative	negative
$x \cdot y$	positive	negative	negative	positive
x / y	positive	negative	negative	positive

For example, when the variables x and y are positive, the sum $x+y$ is positive, the difference $x-y$ is also positive since $|x| > |y|$ in this table, and so on. Before moving on, you should take a minute to look this table over and make sure that you understand all of the different outcomes. If you get confused, just plug in real numbers for x and y.

POP QUIZ:
POSITIVE AND NEGATIVE OUTCOMES . . .
PART TWO

The following table should look pretty familiar, but it's not identical to the previous one . . . there's one big difference. Can you find it? It's right at the top: instead of $|x| > |y|$, this time it's $|x| < |y|$. So, which of the outcomes will change? Think about it, fill in your "positive" or "negative" responses, and then check your answers in the Math Dude's Solutions at the end of the book.

$$|x| < |y|$$

	$x > 0$ $y > 0$	$x > 0$ $y < 0$	$x < 0$ $y > 0$	$x < 0$ $y < 0$
$x + y$	positive			
$x - y$		positive		
$x \cdot y$			negative	
x / y				positive

ARITHMETIC 2.0 (HERE COMES THE ALGEBRA)

What exactly do I mean by arithmetic 2.0? It's simple really. Arithmetic 2.0 is just the normal everyday arithmetic that you know and love (basically arithmetic 1.0) . . . with a few extras: variables, exponents, and roots. Just to be clear, this isn't a technical term or one your teachers will use or anything like that; it's just my way of saying that even though some of the stuff we're going to be doing might look unfamiliar and complicated, at its heart it really is just arithmetic.

ADDING AND SUBTRACTING VARIABLES

We've already spent a lot of time working with variables in previous chapters, so some of the things you'll see here will look familiar. So, what can we do with variables? Well, pretty much anything we can do with numbers! As we've seen, we can add a number to a variable to make an expression

then we can subtract a number from that expression to get a new expression

Or, we could have instead added a variable to our original variable

In other words, everything works in exactly the same way as it does with numbers. Well, almost. The one big difference is that we don't know the actual number of boxes in any of these rows! But that doesn't stop us from being able to do arithmetic with them. And we're not restricted to working with one variable, we can throw in an additional term and add three variables:

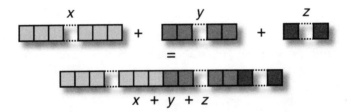

ALGEBRA DECODER!

• TERM

"Let's throw in an additional term?" Uh . . . what? A term? Yes, a term. The easiest way to describe what a **term** is in math is with an example. Take a look at the expression:

$2x^2 - 2x + 4$

There are three terms here: $2x^2$, $2x$, and 4. As you may have guessed, the terms in an expression are all the pieces separated by addition or subtraction (but not multiplication or division). There's not really any math in this term (pardon the pun), the word just gives us a convenient way to talk about the various parts of an expression.

MULTIPLYING VARIABLES

But why stop with addition and subtraction? As we've seen in previous chapters, we can also multiply variables. For example, we can multiply $x \cdot y$ with the result that the magnitude of y is scaled to be x times bigger than its initial size.

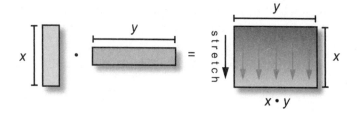

Which means that $x \cdot y = xy$ is just the area of this rectangle. Assuming y is a positive number ($y > 0$), multiplying y by x can have several outcomes . . . all nicely summarized in this table:

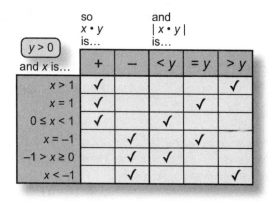

$y > 0$ and x is...	so $x \cdot y$ is... +	−	and $\lvert x \cdot y \rvert$ is... $< y$	$= y$	$> y$
$x > 1$	✓				✓
$x = 1$	✓			✓	
$0 \leq x < 1$	✓		✓		
$x = -1$		✓		✓	
$-1 > x \geq 0$		✓	✓		
$x < -1$		✓			✓

Along the left side you can see the different ranges of values that x can have, and on the right you can see the resulting sign and magnitude of $x \cdot y$. Take a minute to look through these and make sure you understand them.

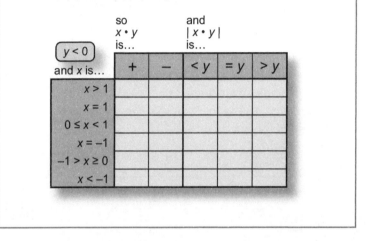

POP QUIZ: WHAT DO YOU GET?

What happens to the sign and magnitude of $x \cdot y$ if we assume that $y < 0$ instead of $y > 0$? Fill in this table with your answers. You might want to stop and think for a minute before you start going through the problem. Is there anything about absolute values that might help you answer the problem more quickly?

$y < 0$ and x is...	so $x \cdot y$ is...		and $\lvert x \cdot y \rvert$ is...		
	$+$	$-$	$< y$	$= y$	$> y$
$x > 1$					
$x = 1$					
$0 \le x < 1$					
$x = -1$					
$-1 > x \ge 0$					
$x < -1$					

What else can we do? Well, when we multiply a number by itself we square the number, and when we multiply a variable by itself we square the variable:

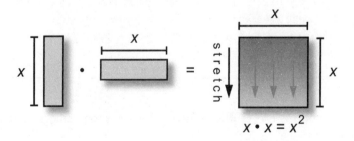

$$x \cdot x = x^2$$

In other words, we stretch out the magnitude of the variable x until it's x times its original size. How about multiplying a variable by itself more than once? Well, here's what you get when you multiply *x* by itself twice:

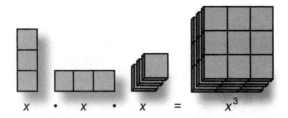

In other words, you can think of x^3 as being made by stretching the square formed by x^2 up out of the page until you've formed a big cube made up of $x \cdot x \cdot x = x^3$ little 1-by-1-by-1 unit cubes. And that means that you can think of x^3 as representing the volume of the big cube!

ALGEBRA DECODER!

• VOLUME

Like area, **volume** is something you use all the time in everyday life. When you buy milk or fill up your gas tank, you're buying a certain volume of liquid (measured in units like gallons and liters). And that makes sense when you look at our picture of a cube. After all, it's a box that you could fill with stuff. How much? Well, if *x* is the size of the length, width, and height of the box, then $x \cdot x \cdot x = x^3$ tells you how much will fit. Which exactly mirrors our everyday interpretation of volume as being length • width • height.

If you instead multiply three different variables *x*, *y*, and *z*, you get

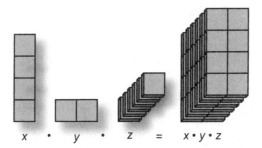

$$x \quad \bullet \quad y \quad \bullet \quad z \quad = \quad x \bullet y \bullet z$$

Again, this means that you can think of the product of the three variables x, y, and z as forming a big box that contains $x \bullet y \bullet z$ little 1-by-1-by-1 unit cubes.

DIVIDING VARIABLES

Finally, just as we've seen that we can stretch variables by multiplying them by other variables, we can also compress variables by dividing them by other variables, like this:

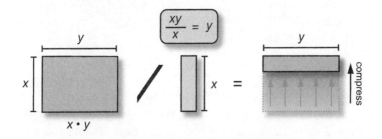

So that is a quick rundown of how variables work in the arithmetic world. Most important, you now have a way to picture the meaning of the four main arithmetic processes . . . which will come in handy as we move on to doing more and more complex things with variables.

HOW TO SIMPLIFY EXPRESSIONS

Okay, let's take a minute to see how we can combine every-thing we've learned so far to help us understand what it means to simplify an expression. And while we're at it, let's go through an easy-to-follow three-step plan to help you sim-plify any expression at all. Let's use this expression:

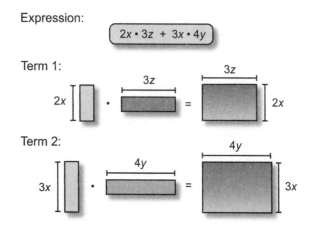

Expression:

$$2x \cdot 3z + 3x \cdot 4y$$

Term 1:

Term 2:

As you can see, this expression has two terms—both of which ultimately can be represented as the area of a rectan-gle. But as it's written, this expression is a little messy. And by that I mean there are a lot of extra numbers floating around just waiting for us to simplify them.

■ STEP 1: WORK TERM BY TERM

Okay, this first step is less of an actual step and more of a gen-eral rule of thumb . . . but it's a useful one. So, what do I mean by working "term by term"? Well, I mean that you should sim-plify each term separately. I don't mean that you can't do them simultaneously, I just mean that you should look at each term as its own little problem that doesn't interact with the other terms.

■ STEP 2: GATHER CONSTANTS AND VARIABLES
IN EACH TERM

The next step in simplifying an expression is to gather together all of the numerical constants and variables in each
term separately. For our problem, this means that we go from

$2x \bullet 3z + 3x \bullet 4y$

and turn it into

$(2 \bullet 3)xz + (3 \bullet 4)xy$

All that we did was put all the constants, or numbers, in
each term inside parentheses, which left all the variables
grouped outside the parentheses.

■ STEP 3: SIMPLIFY EACH TERM

It's time to simplify each term. Do you see why we put all the
constants and variables in their own groups? Because those
are the things that we can simplify! In this problem we can
only simplify the constants at this point, but there are plenty
of problems where you might find yourself multiplying variables together now too (more on that later in the chapter). So,
to finish simplifying our expressions, we just need to multiply some numbers:

$6xz + 12xy$

And that's as simple as we can get this expression. Which
means that we're done! I should mention one piece of termi-
nology that you'll see me using as we move forward (especially
in the next chapter). The numbers 6 and 12 in this expression are known as "numerical coefficients." Numerical coefficients are just the numbers that come in front of the variables
in a term. If you want to understand what these numerical

coefficients really mean in this problem, take a look at this visual representation of the first term

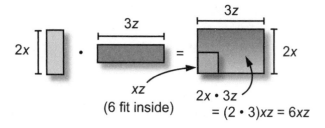

As you can see, the numerical coefficient 6 in this term is just the number of little xz rectangles that fit inside the big rectangle that we get by multiplying $2x \cdot 3z$. Similarly, for the second term

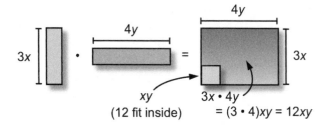

you can see that the numerical coefficient 12 is just the number of little xy rectangles that fit inside the big rectangle that we get by multiplying $3x \cdot 4y$.

To sum up, here are the three steps you should follow when simplifying expressions:

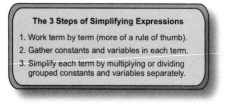

The 3 Steps of Simplifying Expressions

1. Work term by term (more of a rule of thumb).
2. Gather constants and variables in each term.
3. Simplify each term by multiplying or dividing grouped constants and variables separately.

Q&DT ▶

✏ **POP QUIZ:**

LET'S SIMPLIFY SOME EXPRESSIONS!

Use the three-step process to simplify the following expressions:

1. $2x \cdot 3x - (10z / 5z) =$ _____

2. $8z / 2x + 4y =$ _____

3. $7z \cdot 4x / 7y + 3xy / 7x - 9x =$ _____

MATH PROPERTIES

The following properties of arithmetic are most likely almost second nature to you. And by that I mean that you use them when solving math probems without even thinking about the fact that you're using them. While that's just fine, it's interesting to stop and realize that a lot of these properties that you and I take for granted as being "obvious" aren't actually as obvious as we think! Here's a brief rundown of these properties.

THE COMMUTATIVE PROPERTY OF ADDITION

The commutative property of addition says that the order in which you add two numbers or variables together doesn't matter. In other words, $x + y = y + x$:

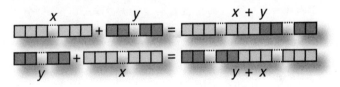

THE ASSOCIATIVE PROPERTY OF ADDITION

The associative property of addition says that the way in which you group numbers or variables when adding three or more together doesn't matter. In other words, $(x+y)+z=x+(y+z)$:

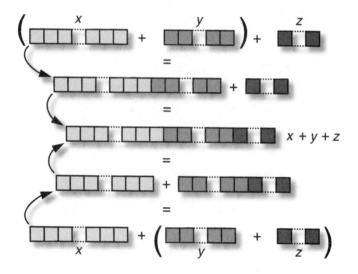

✎ POP QUIZ: HOW TO ADD QUICKLY

How would you go about adding: 2, 7, 9, 8, 1, 2, 3, 3, 5, 9, 1? By simply working from left to right like $2+7=9$, then $9+9=18$, then $18+8=26$, and so on? If so, there's a much faster way (and no, it's not a calculator). The trick is to find pairs or groups of numbers that add to 10 (since it's really easy to add 10s). In other words, instead of doing this

$$2+7+9+8+1+2+3+3+5+9+1$$

you want to do this (eventually in your head):

$$(2+8)+(7+3)+(9+1)+(2+3+5)+(9+1)=10+10+10+10+10=50$$

Cont'd

Cont'd from page 224

How can we just move the numbers around and regroup them like that? Because the commutative and associative properties of addition say we can! Try using this technique to add up the following list of numbers:

3, 9, 2, 6, 4, 2, 7, 5, 1, 9, 3 .... sum =_____

THE COMMUTATIVE PROPERTY OF MULTIPLICATION

The commutative property of multiplication says that the order in which you multiply two numbers or variables doesn't matter. In other words, the two rectangles formed by $x \cdot y$ and $y \cdot x$ have the same area, which means that $x \cdot y = y \cdot x$:

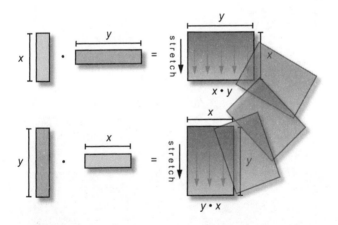

While you may be tempted to think that this property is "obvious," it really isn't. Let's say the two problems are $3 \cdot 5$ and $5 \cdot 3$. The first one means "give me three groups of 5" or "stretch 5 until it's three times its original size," and the second one means "give me five groups of 3" or "stretch 3 until it's five times its original size." Of course we know both answers are 15 . . . but it's not "obvious" that both of these are the same!

THE ASSOCIATIVE PROPERTY OF MULTIPLICATION

The associative property of multiplication says that the way in which you group numbers or variables when multiplying three or more of them doesn't matter. In other words, the volumes of the two cubes formed by $(x \bullet y) \bullet z$ and $x \bullet (y \bullet z)$ are equal, which means that $(x \bullet y) \bullet z = x \bullet (y \bullet z)$:

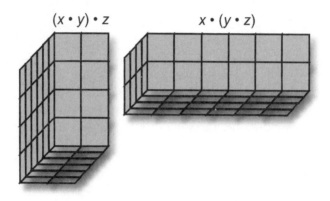

THE DISTRIBUTIVE PROPERTY

So far we haven't mixed multiplication with any of our other arithmetic operations. But that's all about to change. Take a look at this drawing:

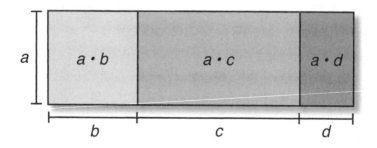

What does it mean? Well, a, b, c, and d here are all variables that describe the lengths of the sides of the three rectangles in the figure. As you can see, the area of the rectangle on the left is equal to $a \cdot b$, the area of the middle rectangle is $a \cdot c$, and the area of the rectangle on the right is $a \cdot d$. So what's the total area of all the rectangles? It must be the sum of these three areas:

$a \cdot b + a \cdot c + a \cdot d$

Okay, but if you put those three rectangles together they form a bigger rectangle. What's the area of that big rectangle? It's just its height, a, times its width—which is $b + c + d$. And that means that the total area of the big rectangle is

$a \cdot (b + c + d)$

But a little earlier we said that the total area of the big rectangle is $a \cdot b + a \cdot c + a \cdot d$. Which, if you think about it, means that

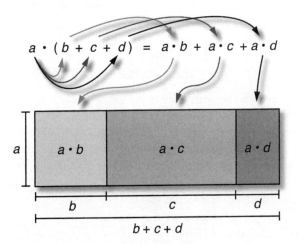

This method of distributing the product of one number or variable (in this case the variable a) over the sum of some other numbers or variables is a perfect demonstration of what's called the distributive property.

Distribute Over Addition and Subtraction Only

Multiplication is *only* distributive over addition and subtraction . . . it's not distributive over multiplication and division! Which means that while it's fine to do things like

$$a \bullet (b+c) = a \bullet b + a \bullet c \quad \checkmark \text{ Right}$$

and

$$a \bullet (b-c) = a \bullet b - a \bullet c \quad \checkmark \text{ Right}$$

it is *not* okay to do things like

$$a \bullet (b \bullet c) = (a \bullet b) \bullet (a \bullet c) \quad \times \text{ Wrong!}$$

and

$$a \bullet (b / c) = (a \bullet b) / (a \bullet c) \quad \times \text{ Wrong!}$$

WATCH OUT!

POP QUIZ: WHICH PROPERTY LETS YOU DO THIS?

You've now learned about the big five properties of arithmetic that describe how we can add, subtract, multiply, and divide numbers and variables in algebra. For each of the following equations, figure out which property has been used to go from the expression on the left to the expression on the right. Your choices are:

- Commutative Property of Addition (CA)
- Commutative Property of Multiplication (CM)
- Associative Property of Addition (AA)
- Associative Property of Multiplication (AM)
- Distributive Property (D)

1. $1 + (2+3+4) = (1+2+3) + 4$ .... _____
2. $15 + 2 + 7 = 15 + 7 + 2$ .... _____

Cont'd

Cont'd from page 228

3. $2 \bullet (4+5) - 1 = 2 \bullet 4 + 2 \bullet 5 - 1$ .... _____
4. $(5 \bullet 3) \bullet 1 + 7 = 5 \bullet (3 \bullet 1) + 7$ .... _____
5. $8 + 3 \bullet 8 = 8 \bullet 3 + 8$ .... _____

One of these five problems actually requires two properties. Can you figure out which?

A PICTURE IS WORTH A SINGLE EXPRESSION

Earlier we took a visual approach to figuring out why the distributive property works. Now, I want to take that one step further and challenge you to figure out which equation this drawing represents:

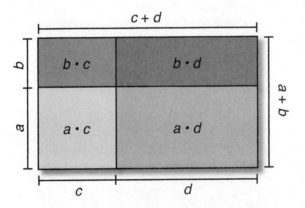

Not surprisingly, the expression you come up with should contain the variables *a, b, c,* and *d*. And it should also use the distributive property of multiplication. Give it some thought, write out your ideas, and then read on for the solution.

FOIL: FIRST, OUTER, INNER, LAST

The first thing to notice about our drawing is the obvious: It's a big rectangle that contains four smaller rectangles. Following in the footsteps of our earlier problem, let's try to come up with expressions for the areas of the smaller rectangles. As you can see in the drawing, starting from the bottom left and working clockwise, the four smaller rectangles have areas of $a \bullet c$, $b \bullet c$, $b \bullet d$, and $a \bullet d$. So that means that the total area of all four rectangles is

$$a \bullet c + b \bullet c + b \bullet d + a \bullet d$$

Or, using the shorter notation for multiplication that leaves out the dots

$$ac + bc + bd + ad$$

Now, just as we did in our previous problem, let's think of a way to write the area of the large rectangle. Well, the area is just the height, $a + b$, times its width, $c + d$, which means that the total area is

$$(a + b) \bullet (c + d)$$

If we set these two expressions for the total area equal to each other, we see that the drawing represents the equation

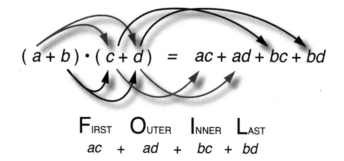

In other words, this equation says that the total area of the rectangle is the same as the sum of the areas of the four smaller rectangles. This is one of those things that is totally obvious when you see the picture, but not so much when looking at the equation. And if you've taken an algebra class before, then I can guarantee you've seen this equation since this is the infamous FOIL formula that tells you how to distribute multiplication over two expressions that each have two terms.

If you're not sure where the acronym FOIL is coming from, take another look at the order of the four multiplications performed on the left side of the equation. FOIL stands for

- The **First** variables from each expression are multiplied and added to the total.
- The **Outer** variables from each expression are multiplied and added to the total.
- The **Inner** variables from the two expressions are multiplied and added to the total.
- The **Last** variables from the two expressions are multiplied and added to the total.

The order of doing the multiplication set by FOIL isn't actually required. You'd get the exact same answer (after rearranging a few terms) by doing something like LIOF instead. But everybody seems to remember it as FOIL . . . probably because it's actually a word.

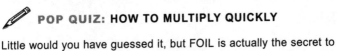 **POP QUIZ: HOW TO MULTIPLY QUICKLY**

Little would you have guessed it, but FOIL is actually the secret to you performing lightning-fast multiplication in your head. Intrigued? Let's think about solving the problem 62 · 27. But instead of writing

Cont'd

Cont'd from page 231

it out and doing the multiplication the old-fashioned way, let's use this

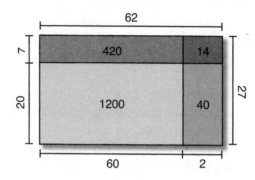

This rectangle gives you a new way to think about multiplying numbers. The first thing you need to do is split each number into two easy to multiply parts (typically you want one part to be a power of 10). In this case, let's split 62 into $60+2$ and 27 into $20+7$. Then the problem becomes

$62 \cdot 27 = (60+2) \cdot (20+7)$

Using FOIL, we get

$62 \cdot 27 = (60 \cdot 20) + (60 \cdot 7) + (2 \cdot 20) + (2 \cdot 7)$

These are all fairly easy to solve in your head, which means you can quickly calculate

$62 \cdot 27 = 1200 + 420 + 40 + 14 = 1674$

Now it's your turn. Use FOIL and the following rectangle to quickly solve $74 \cdot 45$:

Cont'd

Cont'd from page 232

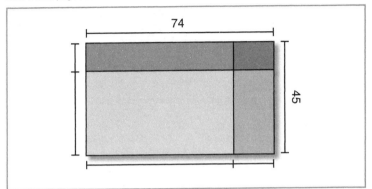

The Disastrous Consequence of Dividing by Zero

Dividing by zero is bad. It causes everything we know and love about math to fall apart. For example, start with the equation

$x = y$

Now add x to both sides and simplify to get

$2x = y + x$

Just for fun, let's now subtract $2y$ from both sides and simplify to get

$2x - 2y = x - y$

Using the distributive property, we can write the left side as $2 \bullet (x - y)$, like this:

$2 \bullet (x - y) = x - y$

Now, since we have $x - y$ on both sides of the equation, we can go ahead and divide both sides by $x - y$. And when we do that we get

$2 = 1$

Oops! That can't be right. But we didn't do anything wrong. Every step was simple, logical, and valid . . . except for the very last

WATCH OUT!

Cont'd

Cont'd from page 233

one. When we divided both sides by $x - y$, we were actually dividing by zero since we started this whole thing out by saying that $x = y$. The bottom line is don't divide by zero . . . or else!

COMBINING LIKE TERMS

Earlier in the chapter we looked at how to simplify expressions by grouping together and then simplifying the numerical constants and variables in each of their terms. But there's another way to make expressions simpler—and that's by doing something called combining like terms.

WHAT ARE LIKE TERMS?

The easiest way to explain what like terms are is with an example. As you know, we can add the variable x to itself to create a new expression $x + x$ that contains two terms, like this:

But we don't have to put the boxes representing the variables x and x end to end like this. In fact, since both variables have the same length, we can stack them one on top of another and create a 2-by-x rectangle, like this:

Does that shape look familiar? It's the shape that's created by stretching the magnitude of x until it's twice its original size. In other words, it's the shape you get by multiplying $2 \cdot x$. And notice too that when we stack the boxes like this, the expression on the right ends up having only one term, 2x, instead of two like before. In other words, because the two variables that we're adding are what are called "like terms," we can combine them into a single term.

What if we had been adding x to y instead of x to x? Would we still have been able to combine the two variables? Well, take a look:

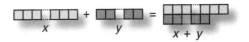

As you can see, the lengths of x and y are different, which means that we can't stack them up and combine them as we did with x + x. And as a result, you can see that the expression on the right still has two terms. In other words, we can't combine x and y into a single term because they are not like terms.

WHAT DOES IT MEAN TO ADD LIKE TERMS?

The first way to combine like terms is to add them. If you've got an x^2 over here, another one over there, and a final one somewhere else, then you can put them all together to get $3x^2$.

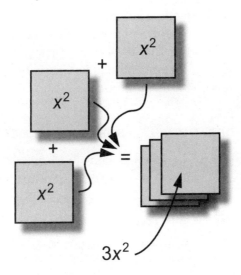

Why can you do that? Because all the x^2 terms are the same type of thing. In our view of the world, this means that they're all of the same shape and size . . . which means that we know how to stack them up into a cube.*

But if you instead have a y^2 here, another y^2 there, and an x^2 way over there, you can't just add them all together since they're not all the same type of object.

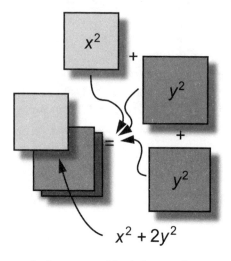

$$x^2 + 2y^2$$

In other words, in our world of shapes this means that since x^2 and y^2 are not the same size, we can't stack them up. Of course, you can stack up and add both of the y^2 terms together since they are the same size and shape. And that's exactly why the final expression for the sum here is equal to $x^2 + 2y^2$.

WHAT DOES IT MEAN TO SUBTRACT LIKE TERMS?

The second way to combine like terms is to subtract them. As you might guess, subtraction is very similar to addition. As long

---

*You'll notice that this "cube" looks a little different than the cubes from earlier in the chapter. From this point forward I'll be using this style of "cube" with stacked sheets instead of boxes since it makes it much easier to see how high the stack is.

as you're dealing with like terms (that is, that the things you're trying to subtract have the same size and shape when represented with boxes, rectangles, and cubes), then you can subtract the numerical coefficients of those terms. For example, since x^2 and $3x^2$ both are some number of x^2, we can subtract x^2 from $3x^2$ to get $2x^2$:

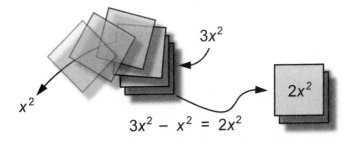

THE BOTTOM LINE

The bottom line is that for two terms to be like, they must have the exact same combination of variables and the exponents that those variables are raised to. In other words, x and x^2 have the same base variable, x, but they are not both raised to the same power and are therefore not like terms. Which, of course, we can also see by noting that x and x^2 have different shapes.

Notice that the numerical coefficients do *not* have to be the same for two terms to be like. So despite the fact that $10xy$ and $5xy$ have different numerical coefficients, they are still like since they have the exact same base variable combination (and thus have the exact same base variable shape). All of that makes perfect sense since it's exactly those numerical coefficients that we add or subtract when combining like terms.

• **ALGEBRA TUTORIAL** •

HOW TO COMBINE LIKE TERMS

Let's now work through the process of combining the like terms in the expression

$$3x^2 - 2y \bullet xy + x \bullet (x + y^2)$$

and simplifying it as much as we can.

■ STEP 1: USE THE DISTRIBUTIVE PROPERTY TO EXPAND EXPRESSION

The first thing to do is to use the distributive property to expand the expression as much as possible. In this case, that means we need to multiply out $x \bullet (x + y^2) = x^2 + xy^2$ in the third term. The newly expanded expression is therefore

$$3x^2 - 2y \bullet xy + x^2 + xy^2$$

■ STEP 2: SIMPLIFY EACH TERM IN THE EXPRESSION

Next we need to simplify each term in the expression. To do this, you can follow the steps outlined in the earlier How to Simplify Expressions tutorial on page 220. Essentially we need to make sure that we've multiplied or divided all the numerical constants and variables in each term until they are as simple as possible. In this case, we can multiply out the product in the second term $2y \bullet xy = 2xy^2$. The newly simplified expression is therefore

$$3x^2 - 2xy^2 + x^2 + xy^2$$

■ STEP 3: MOVE LIKE TERMS INTO GROUPS

The third step is to move all the like terms into groups. Strictly speaking this step isn't necessary, it just helps you to get organized and see where all the like terms are located. In this case, we can group the first and third terms and the second and fourth terms to get

$$(3x^2 + x^2) + (-2xy^2 + xy^2)$$

■ STEP 4: COMBINE THE LIKE TERMS

Finally, it's time to actually combine all of our newly grouped like terms using either addition or subtraction. In our case, we end up with a final expression of

$$4x^2 - xy^2$$

It's certainly a lot simpler than what we started with! To sum up, here's the four-step plan for combining like terms:

The 4 Steps of Combining Like Terms

1. Use distributive property to expand expression.
2. Simplify each term in the expression.
3. Move like terms into groups.
4. Combine the like terms.

◀ Q&DT

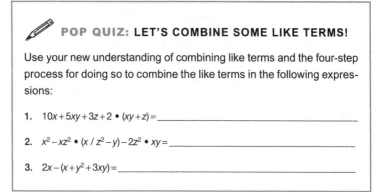

POP QUIZ: LET'S COMBINE SOME LIKE TERMS!

Use your new understanding of combining like terms and the four-step process for doing so to combine the like terms in the following expressions:

1. $10x + 5xy + 3z + 2 \cdot (xy + z) =$ _____

2. $x^2 - xz^2 \cdot (x / z^2 - y) - 2z^2 \cdot xy =$ _____

3. $2x - (x + y^2 + 3xy) =$ _____

EXPONENTIATION

Earlier in the chapter we talked about two ways that you can think about multiplication. One way was to think of multiplication as the process of repeatedly adding the same number to itself—at least when we're talking about multiplying positive integers. In this picture, you can think of $3 \cdot 5$ as $5 + 5 + 5$. Or, more generally, you can think of

$$x \cdot y = \overbrace{y + y + \ldots + y + y}^{x \text{ times}}$$

But what if instead of adding a number or variable to itself several times in a row, we want to multiply it by itself several times in a row? For example, let's say that we want to multiply 5 by itself five times, like this: $5 \cdot 5 \cdot 5 \cdot 5 \cdot 5$. Well, that's definitely a little cumbersome to write. How about multiplying 5 by itself ten times: $5 \cdot 5 \cdot 5 \cdot 5 \cdot 5 \cdot 5 \cdot 5 \cdot 5 \cdot 5 \cdot 5$. Now that's really a pain to write! Clearly we need a way of expressing this idea without having to write so much. Well, we do. It's called exponentiation.

WHAT EXACTLY IS EXPONENTIATION?

We're already very familiar with exponentiation in its simplest form: squaring a number. Anytime we multiply a number or variable by itself, we "square" it—like $x \cdot x = x^2$. We can also "cube" a number or variable—like $x \cdot x \cdot x = x^3$. There aren't any other cute names like "square" and "cube," but we can move on to multiplying a number or variable by itself "n" times (where "n" stands for any number) using the notation x^n . . . usually spoken as "x to the n." For example, when n=4, we say "x to the fourth," "x raised to the fourth power," or "the fourth power of x." They all mean the same thing, although the first is the most commonly used.

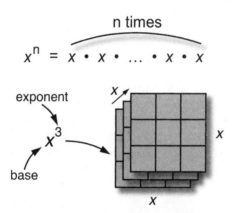

The notation for exponentiation consists of two parts: the first is called the "base," which is the variable x in the example above, and the second is called the "exponent." The exponent is 2 when we square a number, 3 when we cube it, and so on.

HOW TO PICTURE EXPONENTIATION

How should you picture exponentiation? Well, we've already looked at some ways that you can picture what it means to square and cube numbers. Namely, x^2 produces a square and x^3 produces a cube like the one seen above. It gets pretty hard to continue this picture past cubes since we can't visualize things in more than three-dimensions (which is probably why there isn't a special word for powers past "cube"), but you won't see many exponents beyond 3 for a while.

As we found with picturing multiplication as repeated addition, picturing exponentiation as the process of repeatedly multiplying a number by itself becomes hard if the exponent is not a positive integer. For example, while it makes perfect sense that 5^4 can be thought of as meaning $5 \cdot 5 \cdot 5 \cdot 5$, it's not so easy to figure out what $5^{\sqrt{2}}$ is! But luckily, figuring out how to think of exponentiation in these extreme situations won't really matter, for now, so we'll just leave it at that . . . as something you should be aware of, but not worry about.

POP QUIZ: EVEN AND ODD EXPONENTIATION

When you multiply an odd number by an odd number, the answer is always an odd number. And when you multiply two even numbers or an even and an odd number, the answer is always an even number. Can you extend this idea to exponentiation? Specifically, can you figure out a way to tell if a number raised to a power will be even or odd without actually calculating the result? What does the answer depend upon?

HOW TO MULTIPLY WITH EXPONENTS

What does an exponent of 1 mean? Well, the exponent tells you how many of the bases you're multiplying together. For example, x^2 means to multiply two bases of x together to get $x \cdot x = x^2$. And an exponent of 1 therefore tells you to multiply only one base of x. Which means that $x^1 = x$. Okay, now let's see what happens when we multiply $x^1 \cdot x^1$. Well, since $x^1 = x$, that's just the same thing as $x \cdot x = x^2$. How about $x^1 \cdot x^2$? That gives us $x \cdot (x \cdot x) = x^3$. And then $x \cdot x^3$ gives x^4, and so on. We could keep on going, but that's enough for us to figure out the pattern. Here's what we've found:

- $x^1 \bullet x^1 = x \bullet x = x^2$
- $x^1 \bullet x^2 = x \bullet (x \bullet x) = x^3$
- $x^1 \bullet x^3 = x \bullet (x \bullet x \bullet x) = x^4$

The exponent of x in each line is one more than it is in the previous line, which we can summarize like this:

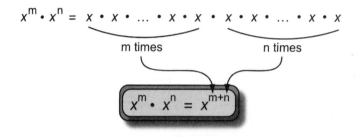

This equation gives us a convenient shortcut for multiplying numbers and variables raised to powers. It says that as long as the bases of the numbers or variables are the same, all we have to do to multiply them is to add their exponents. You can see this equation in action here with some shapes that we're very familiar with.

$x^1 \bullet x^1 = x^2$:

$x^2 \bullet x^1 = x^3$:

Beware of Different Bases when Multiplying with Exponents

Let me emphasize that the bases of the numbers or variables with exponents that you're multiplying *must* be the same in order for you to use this equation. In other words, while $x^m \bullet x^n = x^{m+n}$ works perfectly, you simply cannot do the same thing with two different variables $x^m \bullet y^n$. So if you're trying to simplify the numbers in a problem and you come up against something like $5^2 \bullet 3^3$, don't try to add the exponents!

WATCH OUT!

POP QUIZ:
LET'S MULTIPLY SOME EXPRESSIONS . . . WITH EXPONENTS!

First determine whether or not you can use the equation $x^m \bullet x^n = x^{m+n}$ to help simplify the following problems. If you can, then do it!

1. $5^2 \bullet 3^8 \bullet 5 = $ _____

2. $3^2 \bullet x^2 \bullet x^8 = $ _____

3. $4^2 \bullet 3^5 = $ _____

4. $x \bullet (x^2 + x^3) = $ _____

HOW TO DIVIDE WITH EXPONENTS

We now know what happens to the exponents of variables when those variables are multiplied. But what happens when you divide them? To start us off on our journey toward figuring this out, let's look at the problem:

$x^2 \bullet x^n = x^1$

How can we find the value of the exponent n? Well, we can start by using the equation $x^n \cdot x^m = x^{n+m}$ to multiply the variables on the left. When we do that we get

$$x^{2+n} = x^1$$

Now, the only way for this equation to be true is if the exponents on each side are equal to each other. In other words, $2 + n = 1$ must be true . . . or else the equation itself won't be true. And once you know that, it's easy to solve this equation to find that $n = -1$. Which means that the original equation is

$$x^2 \cdot x^{-1} = x^1$$

But what does a negative exponent like this mean? Well, what other problem do you know that looks like this? In other words, what can you do to x^2 to make it equal to x? You can divide x^2 by x . . . as in x^2/x. And we can put these results together to find that $x^{-1} = 1/x$:

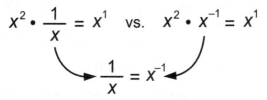

If you wanted to, you could keep doing a similar thing to find that $x^{-2} = 1/x^2$, $x^{-3} = 1/x^3$, and so on. We can combine all these results to write the following equation that you can use to figure out what any negative exponent means:

$$\frac{x^m}{x^n} = x^m \cdot x^{-n} = x^{m+(-n)} = x^{m-n}$$

$$\boxed{\frac{x^m}{x^n} = x^{m-n}}$$

As with multiplication, the bases of the numbers or variables with exponents that you're dividing *must* be the same in order for you to use this equation!

POP QUIZ:

LET'S DIVIDE SOME EXPRESSIONS . . . WITH EXPONENTS!

First determine whether or not you can use the equation $x^m/x^n = x^{m-n}$ to help simplify the following problems. If you can, then do it!

1. $2^2 \cdot 3^8/3^5 =$ _____

2. $x^6/x^3 =$ _____

3. $4^2/3^5 =$ _____

4. $x \cdot (x^2/x^3) =$ _____

WHAT IF ZERO IS AN EXPONENT?

We now know what positive and negative integer exponents are and how they work, but there's one integer that we haven't talked about yet . . . zero. So what does an exponent of zero mean? To find out, let's see what happens when we set m = n in the equation we came up with earlier for the ratio $x^m/x^n = x^{m-n}$. When we do that, we find that

$$\frac{x^m}{x^m} = x^{m-m} = x^0$$

$$1 = x^0$$

So, $x^0 = 1$. In other words, anything raised to the zero power is simply equal to 1. But how can we say that this is true for *anything*? Well, we never specified what x is . . . which means that it could have been *anything!*

WHAT IF AN EXPONENT HAS AN EXPONENT?

So far, so good. We've added, subtracted, multiplied, and divided variables with exponents. But what happens if we raise one of these variables with an exponent to a power? In other words, what if an exponent has an exponent . . . something like $(x^3)^2$? What do we do then? Well, let's go back to thinking about our original interpretation of exponentiation. In that view, an expression like $(x^3)^2$ means multiplying x^3 by itself two times, like this: $x^3 \bullet x^3$. As we discovered when multiplying exponents, this is just equal to $x^{3+3} = x^6$. Which means that we can write the following simple rule to describe what happens when a variable with an exponent is raised to a power:

$$(x^m)^n = x^{m \bullet n}$$

But what if we instead have something like $(xy)^n$. Does the exponent affect both x and y? Well, let's again go back to our original interpretation of exponentiation. In that view, $(xy)^n$ is equal to $xy \bullet xy \bullet \ldots \bullet xy \bullet xy$, where we multiply xy by itself n times. But the commutative and associative properties of multiplication allow us to rearrange the order of the variables to get

$$xy \bullet xy \bullet \ldots \bullet xy \bullet xy = x \bullet x \bullet \ldots \bullet x \bullet x \bullet y \bullet y \bullet \ldots \bullet y \bullet y = x^n \bullet y^n$$

In other words, we see that

$$(xy)^n = x^n \bullet y^n$$

which means that the exponent does indeed affect both variables inside the parentheses.

SUMMARY: PROPERTIES OF EXPONENTS

Here, ready for you to cut out, frame, and hang on your wall, are all the properties of exponents that we've discovered:

Q&DT ▶

Properties of Exponents

1. $x^m \cdot x^n = x^{m+n}$

2. $\dfrac{x^m}{x^n} = x^{m-n}$

3. $(x^m)^n = x^{m \cdot n}$

4. $(xy)^n = x^n \cdot y^n$

✎ **POP QUIZ: ARITHMETIC WITH EXPONENTS**

Simplify the following expressions using the four properties of exponents.

1. $(xy)^2 \cdot x^2/x^3 = $ _____

2. $x \cdot (x^3/x^2)^2 = $ _____

3. $x \cdot [xy - (xy)^2] = $ _____

ROOTS

We now know how to raise *any* number or variable to *any* integer power: be it positive, negative, or zero. But the integers are just a tiny part of the wide world of numbers. Which naturally leads to the question of what happens when we raise a number or variable to a rational or even an irrational power?

WHEN THE EXPONENT IS A FRACTION

Let's start by talking about rational powers—that is, let's look at exponents that are fractions. Specifically, let's look at an expression like $x^{1/2}$. So, what does it mean? Allow me to answer that question with another question that's sure to get your brain moving in the right direction. What is the following expression equal to?

$$(x^2)^{1/2}$$

Now that we've worked out the properties of doing arithmetic with exponents, we can use the third property of exponents to figure out that this expression must equal x since

$$(x^2)^{1/2} = x^{2 \cdot 1/2} = x^1 = x$$

Now, can you think of another way to turn x^2 into x? Well, the square root will do that

$$\sqrt{x^2} = x$$

Which tells us that the exponent $\frac{1}{2}$ means the exact same thing as the square root. So from now on, we can write the square root of a variable as either $\sqrt{x}$ or $x^{1/2}$.

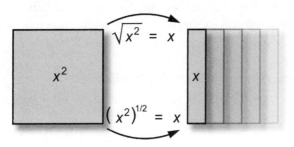

Let's take this one step further and make it a bit more general. Instead of focusing on a specific value of the fractional exponent—such as the $\frac{1}{2}$ in $x^{1/2}$—let's look at the more generic exponent represented by $x^{1/n}$. By following a similar argument that we just made, you can find that

$$(x^n)^{1/n} = x^{n \cdot 1/n} = x^1 = x$$

Which means that the fractional exponent $1/n$ represents what's called the "n-th root."

THE RELATIONSHIP BETWEEN ROOTS AND EXPONENTIATION

We've talked about square roots before, but now it's time to come face-to-face with the general idea of a root. So, what is it? Well, you can think of the root of a number as another number that's used to build that first number by the process of repeated multiplication. That's quite a mouthful, so let's look at a few examples to see what it means:

- $2 \cdot 2 = 4$ means that 2 is the second (aka, square) root of 4. We can write this as either $\sqrt{4} = 2$ or $4^{1/2} = 2$.
- $2 \cdot 2 \cdot 2 = 8$ means that 2 is the third (aka, cube) root of 8. We can write this as either $\sqrt[3]{8} = 3$ (we put the superscript 3 before the radical to show that it's the third root) or $8^{1/3} = 3$.
- $2 \cdot 2 \cdot 2 \cdot 2 = 16$ means that 2 is the fourth root of 16. We can write this as either $\sqrt[4]{16} = 2$ (again, the superscript 4 shows that this is the fourth root) or $16^{1/4} = 2$.

Now let's generalize this idea by looking at

$$y^n = x$$

This equation says that multiplying y together n times gives us x. Now, using what we've learned about roots and exponents, we can raise both sides of this equation to the $1/n$ power. On the left we get

$$(y^n)^{1/n} = y^{n \cdot 1/n} = y^{n/n} = y^1 = y$$

And on the right we get $x^{1/n}$. If we put these two sides together, we get

$$y = x^{1/n}$$

Which tells us that y is the n-th root of x. In other words, plugging $y = x^{1/n}$ back into $y^n = x$, we find that $(x^{1/n})^n = x$. Which means that taking a root "undoes" exponentiation, like this:

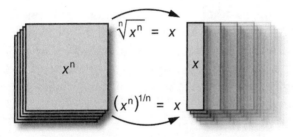

Does that sound familiar? Indeed, in the exact same way that subtraction "undoes" addition, and that division "undoes" multiplication, taking the n-th root is the inverse of the act of raising to the n-th power—a fact that comes in quite handy when solving algebra problems (as we've seen before when taking square roots to solve problems).

✏️ **POP QUIZ: FIND THE EXPONENT**

Can you figure out what the exponent n must be in the following to make this equation true?

$$[(x^2 \cdot 2^3 x)^2]^n = 2x \dots\dots\dots\dots\dots\dots\dots\dots\dots\dots\dots n = \underline{\hphantom{xxxx}}$$

HOW TO DO MATH WITH ROOTS

Do roots play nicely with addition, subtraction, multiplication, and division? In other words, how do you do arithmetic with them? Well, as we've found, roots can be written as exponents . . . so that means that all the rules we've found for dealing with exponents also hold for roots! For example, the fourth property of exponents allows us to figure out how multiplication with roots works:

$$\sqrt[n]{xy} = (xy)^{1/n} = x^{1/n} \cdot y^{1/n} = \sqrt[n]{x} \cdot \sqrt[n]{y}$$

Which means that it's perfectly fine to "break apart" two variables multiplied inside a root. How about division? Well, using the same property, we see that

$$\sqrt[n]{x/y} = (x/y)^{1/n} = (xy^{-1})^{1/n} = x^{1/n} y^{-1/n} = \sqrt[n]{x} / \sqrt[n]{y}$$

Again, we can "break apart" the top and bottom parts of a division problem inside a root. So that means we have quite a bit of freedom when doing multiplication and division with roots. But addition and subtraction are not quite as forgiving. For example, when taking the square root of the sum of two variables, we *cannot* "break apart" the variables like we can when doing multiplication and division. In other words,

$$\sqrt{x+y} \neq \sqrt{x} + \sqrt{y}$$

It just doesn't work! Why? Well, $\sqrt{x+y} = (x+y)^{1/2}$, and there's no rule of exponentiation saying that $(x+y)^{1/2} = x^{1/2} + y^{1/2}$!

For your reference, here's a cheat sheet ready for you to cut out, frame, and hang on your wall with all the properties of roots nicely summarized for you:

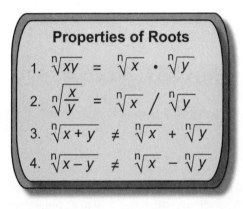

Properties of Roots

1. $\sqrt[n]{xy} = \sqrt[n]{x} \cdot \sqrt[n]{y}$

2. $\sqrt[n]{\dfrac{x}{y}} = \sqrt[n]{x} / \sqrt[n]{y}$

3. $\sqrt[n]{x+y} \neq \sqrt[n]{x} + \sqrt[n]{y}$

4. $\sqrt[n]{x-y} \neq \sqrt[n]{x} - \sqrt[n]{y}$

◀ Q&DT

WATCH OUT!

Don't Trip Over Those Roots

Looking at the third property listed on our roots cheat sheet, I can't help but think back to our old friend the Pythagorean theorem. If we solve $a^2 + b^2 = c^2$ for c, we get

$$c = \sqrt{a^2 + b^2}$$

Upon seeing a square root symbol like this wrapped around those squared values inside, you might be tempted to rush to the conclusion that they "cancel" each other out to give $c = a + b$. But that's just not true—and the third property of roots agrees with me!

✎ **POP QUIZ: ARITHMETIC WITH ROOTS**

Simplify the following expressions using the four properties of roots.

1. $[(x^2y)^2 \cdot x/x^3]^{1/2}$

2. $[x \cdot (x^3/x^2)^2]^{1/3}$

3. $[x \cdot (1-x^2)]^{1/4}$

EXPONENTIATION AND ROOTS COMBINED

So far we've looked at fractional exponents like $\frac{1}{2}$, $\frac{1}{3}$, and even $\frac{1}{n}$. Do you notice anything similar about all of these exponents? Perhaps the fact that they all have a 1 in their numerator? I'd say that's a pretty glaring feature, which might lead you to wonder what a fraction like $\frac{2}{3}$ would mean in an exponent? To figure this out, let's recall our third property of exponents: $(x^m)^n = x^{m \cdot n}$. If we write $\frac{2}{3}$ as the product of 2 and $\frac{1}{3}$, then we see that

$$x^{2/3} = x^{2 \cdot 1/3} = (x^2)^{1/3}$$

And that's all there is to it! Do you see what it means? We can think of a fractional exponent like $\frac{2}{3}$ as having two parts. First, the 2 in the numerator tells us to square x, and then the 3 in the denominator tells us to take the cube root of the result. Or, if we do the same thing with the expression x^3 instead of x, we can picture the process like this:

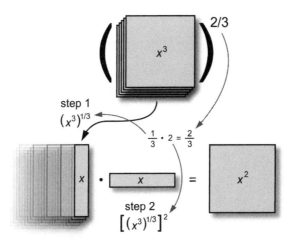

What's going on here? Well, "step 1" in finding $(x^3)^{2/3}$ is to take the cube root of x^3 (that's the $1/3$ part of $2/3$), which has the effect of collapsing the cube into a row. The second step is to square this result (that's the 2 part of $2/3$), which gives us the square, x^2.

Here's how this all works for a more general fraction m/n, where m and n can be any integers:

$$x^{m/n} = (x^m)^{1/n} = \sqrt[n]{x^m}$$

This tells us to first raise x to the power m, and then to find its n-th root. Come to think of it, does it matter which we do first? Do we have to first raise x to the power m and then take the n-th root of the result, or can we instead take the n-th root of x and then raise the result to the power m? Well, multiplication is commutative, which means that $m \cdot \frac{1}{n} = \frac{1}{n} \cdot m$. And therefore

$$x^{m/n} = x^{m \cdot 1/n} = x^{1/n \cdot m} = (x^{1/n})^m = (\sqrt[n]{x})^m$$

What's all that? Look closely and you'll see that it says

$$(x^m)^{1/n} = (x^{1/n})^m \leftrightarrow \sqrt[n]{x^m} = (\sqrt[n]{x})^m$$

which means that the order doesn't matter—we can first raise something to a power and then take the root, or vice versa.

POP QUIZ: **COMBINING ROOTS AND POWERS**

Check and make sure that the order in which you perform the tasks of raising to a power and taking a root really don't matter in the problem $9^{3/2}$.

OVERACHIEVER BADGE

IRRATIONAL EXPONENTS

There's an elephant in the room . . . a big one (are there any small elephants?). Integer exponents make sense, rational exponents make sense, but what in the world could an irrational exponent mean? It can't be a combination of exponentiation and root taking since you can't write an irrational number as a ratio, right? Well, yes that's right. But there actually is a way to think of irrational exponents . . . and it definitely has something to do with exponentiation and root taking. Let me show you what I mean using the irrational number $\sqrt{2} = 1.4142$. First, let's write the decimal representation of $\sqrt{2}$ in the following way:

$$1.4142 \ldots = 1 + \frac{4}{10} + \frac{1}{100} + \frac{4}{1{,}000} + \frac{2}{10{,}000} + \cdots$$

Now, let's use this as the exponent in an expression for $x^{\sqrt{2}}$, like this:

$$x^{1.4142} = x^{1 + \frac{4}{10} + \frac{1}{100} + \frac{4}{1{,}000} + \frac{2}{10{,}000} + \cdots}$$

Cont'd

Cont'd from page 256

This may look kind of strange, but it's an incredibly helpful step for figuring out what irrational exponents mean. Why? Well, all we have to do now is use our first property of exponents, $x^m \cdot x^n = x^{m+n}$, to rewrite the expression for $x^{\sqrt{2}}$ as

$$x^{\sqrt{2}} = x^1 \cdot x^{4/10} \cdot x^{1/100} \cdot x^{4/1,000} \cdot x^{2/10,000} \cdot \ldots$$

What does this mean? It tells us that you can think of an irrational exponent as a list of products, each of which has a rational exponent (which we know how to deal with). The final value, which we can never quite get to since there are an infinite number of products, is equal to the product of all these values. Each successive product going out to the right gets closer and closer to the number 1, and therefore represents a smaller and smaller "correction" for the products to the left of it.

- -
SECRET AGENT MATH-LIB
- -

WHO CARVED THE MYSTERY MESSAGE?

Good afternoon, secret agent _____ *(name of fruit or vegetable)*. Now that you've found your way to work, it's time for lunch! That's right, your cautiously circuitous yet safe route took a big chunk out of your morning, and now you're hungry. You have a real hankering for pasta with _____ *(favorite color)* sauce, and you know that the best place to get that is across the street at The Hungry _____ *(name of zoo resident)* Diner. So, you walk to the restaurant, grab a seat at the counter, and order your lunch.

Moments after the waiter takes your order, you receive a message on your phone. When you look at it, you see something totally bizarre. It looks like some sort of message carved into a stone wall! You can barely make it out, but after a little photo enhancement using a few secret agent math algorithms, you see that it says:

Ahoy!
To ye who decrypts this message, a marvelous bounty awaits.
What's the reward, eh? You'll find out who carved this
message.
So, what're ye wait'n for? Get on with it!

$$\overline{\qquad}_{(1)} \quad \overline{\qquad}_{(2)} \quad \overline{\qquad}_{(3)} \quad \overline{\qquad}_{(4)} \quad \overline{\qquad}_{(5)} \quad \overline{\qquad}_{(6)}$$

And, of course, this message is followed by an algebra problem. Being quite familiar with this routine by now, you waste no time and set out to solve the problem and decode the secret message. After all, you're pretty curious to find out who wrote the message. Was it a pirate?

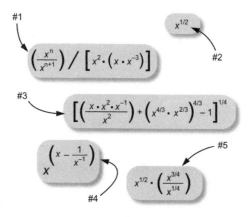

#1 $\left(\dfrac{x^n}{x^{n+1}}\right) / \left[x^2 \cdot \left(x \cdot x^{-3}\right)\right]$

#2 $x^{1/2}$

#3 $\left[\left(\dfrac{x \cdot x^2 \cdot x^{-1}}{x^2}\right) + \left(x^{4/3} \cdot x^{2/3}\right)^{4/3} - 1\right]^{1/4}$

#4 $\dfrac{\left(x - \dfrac{1}{x^{-1}}\right)}{x}$

#5 $x^{1/2} \cdot \left(\dfrac{x^{3/4}}{x^{1/4}}\right)$

The instructions carved into the wall tell you to figure out which of these five expressions goes with each of the following five questions, then to write the number of the expression (one through five) on the first blank line to the right of each question, and finally to perform the indicated arithmetic and write the answer after the equals sign. Good luck!

1. Which expression is equal to x? .... ____ • 5 = ____

2. Which expression will be odd for any value of x? .... ____ + 11 = ____

3. Which expression is equal to x^{-1}? .... ____ + 20 = ____

4. Which expression requires that $x > 0$? .... _2_ • 2 = _4_

5. Which expression is equal to $\sqrt[3]{x^2}$ .... ____ • 3 = ____

Now, for each question, all you have to do is fill in the letter of the alphabet that corresponds to the final answer. Question four is already solved for you. The final answer is 4, which corresponds to "D," the fourth letter of the alphabet. Fill in the rest of the letters to find out who carved that mysterious message!

$$\underline{\hspace{1.5cm}}_{(1)} \quad \underline{\hspace{1.5cm}}_{(2)} \quad \underline{\hspace{1.5cm}}_{(3)} \quad \underline{\hspace{1cm}D\hspace{0.5cm}}_{(4)} \quad \underline{\hspace{1.5cm}}_{(5)} \quad \underline{\hspace{1cm}D\hspace{0.5cm}}_{(4)}$$

Surprise! That's a little odd, isn't it? What's going on? You'll have to keep reading to find out.

CAN YOU REMIND ME OF THE POINT OF ALGEBRA?

Perhaps you're thinking: "Wow, we're sure doing a lot of work with abstract symbols. Is that all that algebra is?" In one sense the answer is, "Yes, that really is it." Because algebra really is all about the logic of working with variables . . . just as arithmetic is about the logic of working with numbers. But in another sense, the answer is no, because after we understand this logic, we can use it to do things like send rockets into space, program computers, and run companies. In other words, after we're fluent in algebra, we can take it to the streets to solve problems.

But we're not quite there yet—there are still a few more things we need to learn first. What's next? Well, it's time for us to learn about the final ingredient needed to tackle the real-world "challenge problems" we talked about at the end of the first chapter. And that ingredient would be? They're called polynomials.

"Excuse me? Poly-what?-ials," you say. Well, welcome to part III . . . coming up right after your chapter 4 final exam.

· FINAL EXAM ·

• ARITHMETIC 1.0: INTEGERS AND FRACTIONS

1. In this chapter, we saw that we could multiply two fractions using the formula

$$\frac{a}{b} \cdot \frac{c}{d} = \frac{a\,c}{b\,d}$$

While that works fine, it's interesting to think about what this means visually too. So, here's a picture for you to think about:

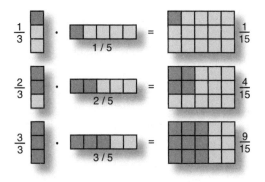

Now, all you have to do is figure out how to finish off these two similar problems. Shade the squares on the right and write down the fractions.

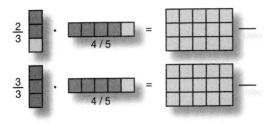

And remember, the real point of this little puzzle isn't to see if you can multiply fractions, it's to get you to look at the relationship between the numerators and denominators of the numbers you're multiplying and those of the result!

- **ARITHMETIC 2.0: VARIABLES AND MATH PROPERTIES**

2. The commutative properties of addition and multiplication are both about the idea that you can switch the order in which you do things without changing the result. Can you think of a real-life example of something that satisfies the commutative property?

3. The associative properties of addition and multiplication are both about the idea that changing the order in which you group things does not change the result. Can you think of a real-life example of something that satisfies the associative property?

4. The distributive property says that $a \bullet (b+c) = ab + ac$. Can you think of a real-life example of something that satisfies this property?

- **EXPONENTIATION**

5. It's common to use something called "scientific notation" to represent numbers (particularly when those numbers are either very large or very small). This notation uses the fact that any number can be expressed as the product of some number and a power of 10, like this:

- $0.0000000065 = 6.5 \times 10^{-9}$
- $0.000003 = 3 \times 10^{-6}$
- $0.001 = 10^{-3}$
- $0.8 = 8 \times 10^{-1}$
- $1 = 1$ (which is actually 10^0)
- $10 = 10^1$
- $3,500 = 3.5 \times 10^3$
- $1,000,000 = 10^6$
- $2,000,000,000 = 2 \times 10^9$

The beauty of using this notation is that it makes doing arithmetic easy. Why? Because, as you now know, it's easy to multiply exponents . . . you just add them. Here are a few problems to help you get the idea:

- First write the number 2,000 in scientific notation (you can use the example of 3,000 above to help)...........................

  _____

- Next write the number 2,000,000 in scientific notation ....

  _____

- Now multiply these together .... _____
- And finally divide them .... _____

Be sure to check the solution at the end of the book if you get stuck!

- TAKING ROOTS . . . SQUARE, CUBE, AND OTHERWISE

6. Try using the expression we found for $x^{\sqrt{2}}$

$$x^{\sqrt{2}} = x^1 \cdot x^{4/10} \cdot x^{1/100} \cdot x^{4/1,000} \cdot x^{2/10,000} \cdot \ldots$$

to estimate the value of $2^{\sqrt{2}}$. How good is the answer you get using just the first two products, $x^1 \cdot x^{4/10}$, versus the first three products, $x^1 \cdot x^{4/10} \cdot x^{1/100}$, the first four, and so on. You can use a calculator to compute the "true" value of $2^{\sqrt{2}}$ (although it's still only an approximation since it's an irrational number).

PART III

SOLVING ALGEBRA PROBLEMS

Where we take everything we've learned so far, add in a few additional ingredients, and use the resulting concoction to solve what Isaac Newton dubbed "universal arithmetic" problems . . . aka algebra problems.

Equations are the devil's sentences.

—Stephen Colbert, *The Colbert Report*

CHAPTER 5

Polynomials, Functions, and Beyond

> Mathematics is the art of giving the same name to different things.
>
> —J. H. Poincaré

We've put together quite an impressive bag of tools for dealing with variables. In fact, at this point we can pretty much do whatever we want with them . . . the mathematical world is our oyster. So that means it's time to start using our tools to build things. In particular, let's use our tools to start building what are called polynomials. That includes monomials, binomials, trinomials, and all the other "nomials" in the world too. So, where to begin? How about with an explanation of the lingo.

In this chapter we'll be talking about lots of math concepts that you'll see not only in math class, but probably on those dreaded SATs too. We'll talk about how to understand polynomials and functions, what these things actually look like and how they work, and we'll also talk about lines, slope-intercept form, solving systems of equations, and all that good stuff!

WHAT ARE POLYNOMIALS?

Polynomials are algebraic expressions that are made up of many terms (the Greek root *poly* means "many"). But those terms can't just be made from whatever you want. There are some rules and restrictions about what exactly polynomials can be built out of.

HOW TO BUILD A POLYNOMIAL FROM SCRATCH

So, how exactly are polynomials built? Well, it all starts with a term. Remember, a term just refers to the product of one or more numbers and variables. For example, $3xy$, $x^{2/3}$, $4xyz$, and $\%$ are all terms. But you'll never see a few of these terms in a polynomial. Why? Because, as I've alluded to before, the terms in polynomials can't just be any old expression. Only particular ones will do. To find out which, let's build a polynomial.

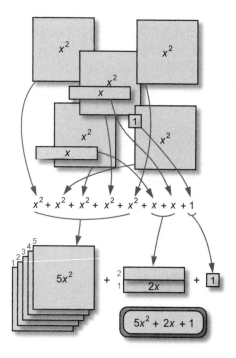

At the top of this drawing you can see some examples of the building blocks from which polynomials are made. What are they? Two things:

1. Numerical constants
2. Variables raised to positive integer powers and *only* positive integer powers

Which means that variables in polynomials can never be raised to a fractional power like $x^{2/3}$ or a negative power like x^{-1}.

As shown in the drawing, you can build a polynomial by adding or subtracting one or more polynomial building blocks. In this case, we've added up a bunch of x^2, a few x, and the numerical constant 1 to create the polynomial expression

$$x^2 + x^2 + x^2 + x^2 + x^2 + x + x + 1$$

That's a pretty big polynomial—it has eight terms! But you may have noticed that many of these terms are identical. Which means that we can combine them.

COMBINING LIKE TERMS IN POLYNOMIALS

In the polynomial we've built, there are five square-shaped x^2 terms and two rectangle-shaped x terms. As we learned in the last chapter, and as you can see at the bottom of the figure, we can combine the identically shaped x^2 terms and x terms into two separate five- and two-stack-high piles. Which means that we can simplify our eight-term expression by combining these like terms to construct the equivalent and much simpler three-term polynomial

$$5x^2 + 2x + 1$$

As is typical for polynomials, each of these terms contains a different power of the variable x.

POP QUIZ:

COMBINING LIKE TERMS OF POLYNOMIALS

Simplify each polynomial as much as possible by combining like terms to create a new expression:

1. $x + 1 + 3x^2 + 3x + 4 =$ _____

2. $x \bullet (2x - x^2) + x^3 =$ _____

3. $x^4 + 3(x^2 - x^4 + 1) - 2 =$ _____

THE ANATOMY OF A POLYNOMIAL

Now that we've seen a few examples of polynomials, it's time to take a more detailed look at their anatomy and the language we use to describe them. The following polynomial contains three terms. The first two are made from constant numerical coefficients and variables (raised to different powers), and the final "constant term" contains only a numerical constant.

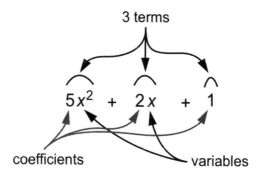

As always, terms can be separated by either addition or subtraction. And, of course, each of the terms can be much more complex than the simple examples seen here. Finally, since the exponents of variables in polynomials can only be positive integers, if you see an expression with a negative, fractional, or irrational number as an exponent, you know that you're not looking at a polynomial.

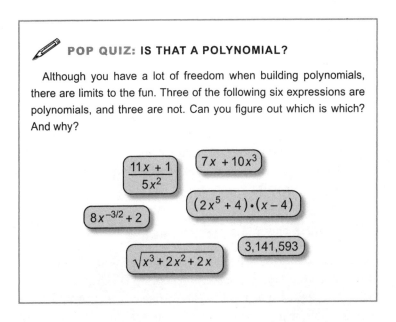

POP QUIZ: IS THAT A POLYNOMIAL?

Although you have a lot of freedom when building polynomials, there are limits to the fun. Three of the following six expressions are polynomials, and three are not. Can you figure out which is which? And why?

$$\dfrac{11x + 1}{5x^2}$$

$$7x + 10x^3$$

$$\left(2x^5 + 4\right) \cdot \left(x - 4\right)$$

$$8x^{-3/2} + 2$$

$$\sqrt{x^3 + 2x^2 + 2x}$$

$$3,141,593$$

MONOMIALS, BINOMIALS, AND TRINOMIALS

There are a few different flavors of polynomials that show up frequently, so they've been given special names to make referring to them easier. The first of these, monomials, are polynomials that have only one term left after combining all like terms. For example, 1 is a monomial. So are x, $3x^2$, $-5x$, and so on. Similarly, binomials are polynomials with two terms left after combining all like terms. Which means that $2+x$, x^2-x^3, $3x^2+x^2+1$ (once we combine like terms), and so on, are all binomials. And, not surprisingly, trinomials are polynomials with three terms.

To help you keep all these terms straight, and to introduce you to a new one—the degree of a polynomial—here's a table giving an example of each type of polynomial we've talked about:

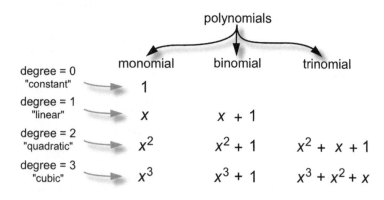

Of course there are an infinite number of possible monomials, binomials, and trinomials that I could have listed here. So instead of picking random ones, I've given one of the simplest possible polynomials in each category.

THE DEGREE OF A POLYNOMIAL

Notice that on the left side of the table above I've indicated something called the "degree" of the polynomial. The degree of a polynomial is the highest power of the variable that it contains. For example, all the polynomials with a degree of 3 shown on the last row have a term that contain the third power of the variable x. That is, they all have x^3 in them. Similarly, polynomials with a degree of 2 all contain x^2, and those with a degree of 1 all contain $x^1 = x$.

And what about that lonely monomial with a degree of 0. Why is it zero? Because $x^0 = 1$. Which means that we could have written the expression 1 as $1x^0$ or even just as x^0 instead. Its meaning would have been exactly the same since x^0 is just another way to write 1.

There are a few other words that are frequently used to describe the degree of a polynomial. As you can see in the table, a degree 0 polynomial is called a "constant" since its value doesn't change (it has no variables), a degree 1 polynomial is called "linear" since they're what the linear equations we looked at in chapter 3 contain, a degree 2 polynomial is called "quadratic," and a degree 3 polynomial is called "cubic."

But why does the degree even matter? It seems like a weird thing to make a big deal over, right? Rest assured that this isn't just hype—the degree of a polynomial is a huge deal. In fact, I can't overstate just how big of a deal it is. As we'll see later, the degree has a dramatic effect on the behavior of polynomials and on how they're used to solve problems.

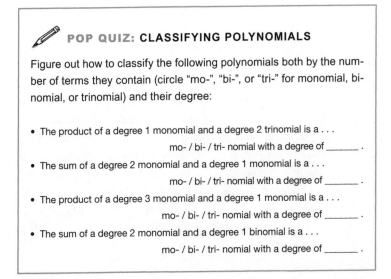

POP QUIZ: **CLASSIFYING POLYNOMIALS**

Figure out how to classify the following polynomials both by the number of terms they contain (circle "mo-", "bi-", or "tri-" for monomial, binomial, or trinomial) and their degree:

- The product of a degree 1 monomial and a degree 2 trinomial is a . . .

  mo- / bi- / tri- nomial with a degree of _____ .

- The sum of a degree 2 monomial and a degree 1 monomial is a . . .

  mo- / bi- / tri- nomial with a degree of _____ .

- The product of a degree 3 monomial and a degree 1 monomial is a . . .

  mo- / bi- / tri- nomial with a degree of _____ .

- The sum of a degree 2 monomial and a degree 1 binomial is a . . .

  mo- / bi- / tri- nomial with a degree of _____ .

WHY POLYNOMIALS?

Now, the 10^{100} (called a googol . . . no, not a Google) dollar question: Why have we bothered to build polynomials at all? And why are there all of these complicated rules and requirements like no negative or fractional exponents? Well, the reason is simple: It's because our goal all along was to construct expressions that look like this

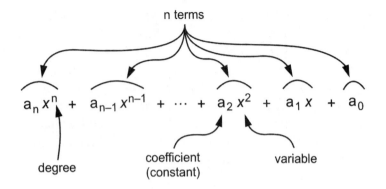

What exactly is *this*? It's a general way to write what we mean by a polynomial. In other words, it's an expression with some number of terms (which we've called "n"), each of which is the product of a constant coefficient and a variable raised to some positive integer powers.

Okay, but why do we want to build expressions that look like that in the first place? Because they play a critical role in a surprisingly large number of problems in math and science. In fact, remember way back at the end of chapter 1 when I talked about three "challenge problems" that we couldn't yet solve, but that we'd be able to solve by the time we got to the end of the book? Well, the reason we couldn't solve them at that point was because we didn't know about polynomials!

✏ POP QUIZ: STANDARD FORM FOR POLYNOMIALS

In math, as in life, it's a good idea to keep things organized. Which is exactly why there's something called a "standard form" that you should use to write polynomials and keep things nice and tidy. In standard form, polynomials are always written with the highest power exponent in the first—also known as the "leading"—term. The term with the next highest power is written to the right of that, and so on, until finally putting the constant term (if there is one) on the far right.

Cont'd

Cont'd from page 272

For example, $x^2 + 2x + 1$ is in standard form since the term containing x^2 is in front, followed by the term containing x, and then the constant term. Now it's your turn. Rewrite the following polynomials to put them in standard form:

1. $2 + 5x^2 - 3x =$ .... _____
2. $x^4 - 10 + x^3 + 5 =$ .... _____
3. $x + x^2 + 1 =$ .... _____
4. $x^2 - x + 2x^2 =$ .... _____

OVERACHIEVER  BADGE

IS EVERYTHING A POLYNOMIAL?

So as not to give you the impression that absolutely everything in algebra is based on polynomials, there are indeed many expressions in the world that are *not* polynomials. And there are many problems in math and science that use them.

RATIONAL EXPRESSIONS

For example, many problems can be written in terms of something called a rational expression—which is just a fancy word for an expression made from the ratio of two polynomials. In other words, a rational expression is something like

$$\frac{x^2 - 2x + 3}{x^3 + 3x^2 - 1}$$

where both the numerator and denominator are themselves polynomials. Of course, the whole thing isn't a polynomial since there are powers of x in the denominator.

Cont'd

Cont'd from page 273

If these other things exist, then why are we focusing so much time on polynomials? Well, for one thing, polynomials are relatively simple and easy to work with—which makes them great to use when you're learning algebra. And second, while not every single problem in the world can be solved using polynomials, a *lot* of them can be . . . including the "challenge problems" that we posed at the end of chapter 1.

OTHER EXPRESSIONS

Is every expression in the world of algebra either a polynomial or a rational expression? No, there are other types of expressions too. In fact, we already know that this has to be true since polynomials, or the ratio of two polynomials, are not allowed to have fractional powers like square roots. And we're well acquainted with a very famous equation that contains a square root: the Pythagorean theorem.

EVALUATING POLYNOMIALS

Now that we know what polynomials are, let's begin our journey toward the magical land where we actually use them to solve problems. And the first step for us in that journey is to think about what it means to evaluate a polynomial. Actually, that might be a better second step—because we'd better first figure out what I mean by "evaluate." No, it doesn't mean that we're going to figure out whether or not your polynomial is doing a good job at school or work. It means we're going to finally plug some actual numbers in for our variables and see what we get back.

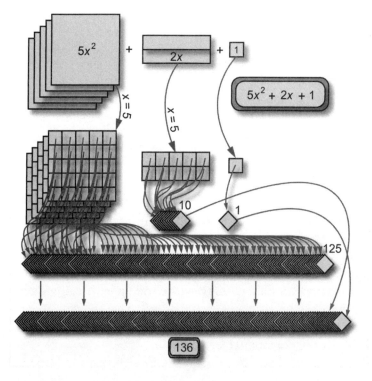

Let's begin by looking at what this all means in terms of the cubes, squares, and rectangles that we first saw in the last chapter. Along the top of the drawing above you'll see a familiar set of shapes representing the polynomial $5x^2 + 2x + 1$. But notice that there's no way to tell from these shapes how big they are. And that's precisely because until we assign a value for the variable x, they don't have any size at all! After all, x could be any number—positive, negative, integer, fraction, irrational, or whatever . . . we just don't know.

But something pretty amazing happens as soon as we assign a value for the variable: all of a sudden, we know exactly how big these shapes are! Let's say we set the variable to $x = 5$. You probably realize by now that we evaluate the expression $5x^2 + 2x + 1$

at $x = 5$ simply by plugging the number 5 in wherever you see an x, like this:

$$5 \cdot 5^2 + 2 \cdot 5 + 1 = 125 + 10 + 1 = 136$$

But to illustrate what that really means, we can draw little 1-by-1 unit squares inside the shapes in our drawing above . . . since once we set $x = 5$, the shapes have an actual size and we now know how big a 1-by-1 unit square should be.

Now imagine pulling apart each of the little unit squares from those shapes and stacking them up. The cube from the first term collapses into a pile of 125 unit squares, the rectangle from the second term collapses into a pile of 10 unit squares, and the lone square from the constant term gives us a third pile with a single unit square. If we then take each of these piles of unit squares and stack them together, we end up with a single giant stack of 136 unit squares. You may not have thought about it this way before, but that process is exactly what it means to evaluate the polynomial $5x^2 + 2x + 1$.

Of course, we don't normally go through that whole mental process each and every time we want to evaluate an expression. That is definitely not the best way to actually solve algebra problems. The most efficient and best way is what we did first—simply plugging in $x = 5$.

Q&DT ▶ But whenever you're feeling like algebra is nothing but a bunch of abstract and hard-to-picture processes, remember that you can always think about it in terms of very real cubes, squares, and rectangles.

POP QUIZ: EVALUATING POLYNOMIALS

Evaluate the following polynomials for the indicated values of x:

1. $5x^2 - 3x + 2$ at $x = 3$ evaluates to .... _____

2. $x^4 + x^3 - 5$ at $x = 2$ evaluates to .... _____

3. $x^2 + x + 1$ at $x = 4$ evaluates to .... _____

4. $3x^2 - x$ at $x = 5$ evaluates to .... _____

OVERACHIEVER BADGE

MULTIVARIATE POLYNOMIALS

While we've been focusing on polynomials that contain only one variable, there's nothing stopping us from building polynomials with two, three, four, five, six, or however many variables we want. As long as we follow the rules we've set out earlier (like no negative or noninteger exponents on any variable), it's a polynomial. In fact, it's what's called a "multivariate polynomial." For example, here's a three variable trinomial: $x^2y - z^3 + 2x^2$.

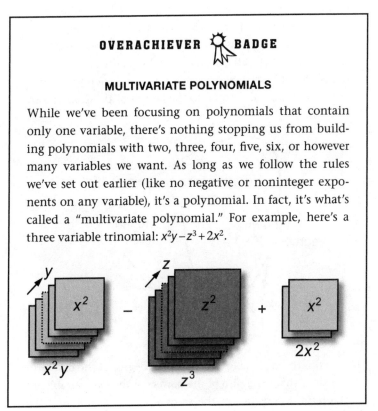

Cont'd

Cont'd from page 277

Combining like terms with multiple variables is no different than combining like terms with one variable . . . it's just slightly more complicated. For example, in the drawing of our three variable trinomial above, we have different sets of cubes and squares for the three unique combinations of variables that appear in the expressions x^2y, z^3, and x^2.

But what's the degree of a polynomial like $x^2y - z^3 + 2x^2$? With a single variable it's pretty easy—the degree is just the highest power of the variable. But now we have more than one variable: y is raised to the first power, x is raised to the second power, and z is raised to the third power. So what's the degree? Well, all you have to do is add up the exponents of all the variables in each term, and then the degree of the polynomial is equal to the largest of these. Here's an example:

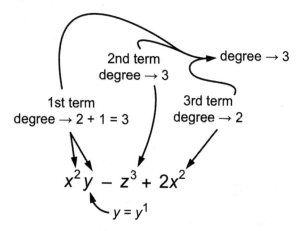

In our three variable trinomial, the degree of the first term is 3 since that's the sum of the powers of the two variables in the term, the degree of the second term is also 3, and the degree of the third term is 2. The degree of the polynomial is the largest of these three values. In other words, it's 3.

FUNCTIONS

Now that we've built some polynomials and figured out what it means to evaluate them, all that's left to do is to actually use them to help us solve problems! And to do that we're going to make a leap out of the world of expression and equations, and into the world of functions.

FROM EQUATION TO FUNCTION

As with any expression in algebra, we can put polynomials to use by creating equations with them. For example, near the end of chapter 3 we solved a problem about finding two numbers whose mean value is one greater than their range. Remember that? Well, little did we know at the time, but the equation we ended up writing to solve that problem included something that we now know to be a polynomial:

$y = 3x - 2$

In fact, we now know that the expression on the right side of this equation is more specifically a degree 1 binomial. The interesting thing about this equation is that unlike most of the equations we've solved in this book, this one has two variables: x and y. What does that mean?

Well, we've found that we can think of a single variable equation like $3x - 6 = x$ as a scale that tests to see if both sides of the equation are balanced.

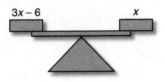

In this case, when $x = 3$ both sides of the scale equal 3—which means that it's balanced. And once you understand this about the meaning of single variable equations, you also understand what equations with two or more variables mean.

So, just as we can think of a single variable equation as a scale, we can also think of a two variable equation like $y = 3x - 2$ as a scale that tests to see if both sides are balanced.

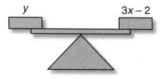

The big difference here is that there isn't just one number to plug into a variable on this scale to check if the equation is balanced. Instead, we can use the fact that $y = 3x - 2$ to calculate any of an infinite number of pairs of numbers that will balance the scale. For example, if we set $x = 1$, then we find that $y = 3 \cdot 1 - 2 = 1$. If we then plug this pair of x and y values into the equation on our scale, you'll see that it's balanced since both sides equal 1.

Okay, that's all well and good, but there's actually another way to think about an equation like this with two variables. And, somewhat surprisingly, that way is to not think about it as an equation at all! So, instead of writing the equation $y = 3x - 2$, let's now investigate what it means to write

$$y(x) = 3x - 2$$

What is that? It's a function.

HOW DO FUNCTIONS WORK?

So, what's different about this? Well, the main outward difference is that we've changed $y \rightarrow y(x)$. But there's more to this change than meets the eye (literally) because this change means that y is no longer just a variable. We're now thinking of y as what's called a function. So what's a function? Well, you can think of a function as a machine that takes some input that you give it and in return gives you back one piece of output. The in-

Q&DT ▸

put to the function is the variable x, which is why it appears in parentheses. Here's how it works:

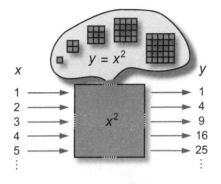

This illustration shows the function $y(x) = x^2$. On the left, you can see a number of different values of the input variable x. Remember, the x in parentheses tells us that x is the input to the function. As you can see, when we feed positive integers such as 1, 2, 3, 4, and 5 as input values of x into the function, the output values that we get (aka the values of y) are simply the squares of the input values. In other words, this is the function that squares numbers!

And just like when we evaluated polynomials earlier, when we feed our function an actual number, we turn the ambiguously sized square representing x^2 into a 1-by-1, 2-by-2, or any other sized square containing lots of small unit squares (you see these in the thought bubble above the big function box). So you can think of the function as something that takes your input, does this calculation, counts up all the little unit boxes, and finally spits out the answer for each piece of input you give it. In other words, the function evaluates the expression.

HOW TO PICTURE FUNCTIONS

It's nice to plug a few numbers into a function and see what output you get like this, but it would be a lot easier to understand what a function does if we could see all of the output from every possible piece of input at once. One way to do this is to feed the function numbers one by one and use the results to make a table. For example, if we do this for the functions $y(x)=x$ and $y(x)=x^2$, we get

x	y(x) = x	y(x) = x²
0	0	0
1	1	1
2	2	4
3	3	9
4	4	16
5	5	25
6	6	36
7	7	49
8	8	64
9	9	81

If we're really ambitious, we could keep doing this on up to higher and higher values of x . . . or maybe we could start plugging in negative values of x. Though that effort would be noble, it doesn't really help us understand the big picture. In other words, it's hard to look at a table and get a feel for the behavior of the function. What we really want is a graph, like this

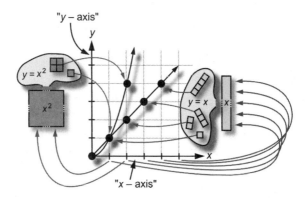

This plot shows the behavior of the two functions $y(x)=x$ and $y(x)=x^2$. We can see input values of x going into each function, and output values of y coming out. After we plot a bunch of these points showing the output for a given input, we can

connect the individual points for each function with curves so that we can see the output values for all the "in-between" input values of x. That means that our plot has given us a picture of the overall behavior of the function. We'll work more on plotting functions a little later in this chapter.

FUNCTIONS VS. EQUATIONS

If you're still a little unclear about the difference between an equation and a function (which is understandable since they do look similar), here's another way for you to think about them. An equation is used to check if stuff on both sides of an equals sign are balanced . . . that's exactly what "equality" means. But a function is instead something that we can plug input into and in return get output. Notice that when it comes to describing what a function is, we haven't said a word about testing for equality.

WHAT MAKES A FUNCTION A FUNCTION?

There's another big difference between a function and an equation: a function *cannot* have more than one y value for the same x value. In other words, in a function like $y(x) = x^2$, every value that you plug in for x gives you back only one value. This is easiest to see by looking at two plots—one of these is a function, and the other is not. Can you figure out which is which?

▸◀ Q&DT

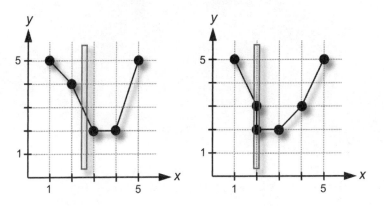

If you look at the graph on the left, you'll see that each point along the x-direction corresponds to only a single value in the y-direction. To see this, you can imagine passing a vertical line back and forth over the graph to check if there are any points along the x-direction that have more than one y-value. Since there aren't, we know that this relationship is a function. But if you do this same test over the graph on the right, you'll find that the point at $x = 2$ has many possible y-values—all of those points on the line between the ordered pairs (2, 2) and (2, 3) have different y-values for the same x-value! And that means that this relationship is *not* a function.

But why can't functions have more than one y-value for the same x-value? The reason is actually pretty simple. The original idea of a function was to be something that we could plug in input and in return get back a *single* piece of output. But in the case of functions, multiple y-values at the same x-value would mean that the function produced more than one output for the same input. And that wouldn't be a function!

POP QUIZ: **IS THAT A FUNCTION?**

Use the vertical line test to figure out which of the following relationships are functions.

Cont'd

Cont'd from page 284

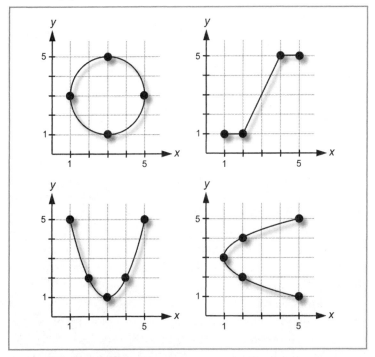

DOMAIN AND RANGE OF FUNCTIONS

As it turns out, we can't just feed a function whatever input we want, we can only give it input that's in what's called its domain. Why? And what does that mean? Well, the domain of a function is simply a list or a description that gives all the x-values at which the function is defined. Let's see how this works for the following functions:

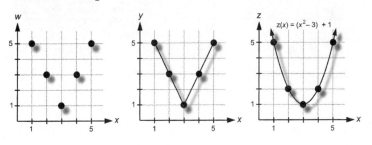

The function on the left contains five values that can be described by the ordered pairs: (1, 5), (2, 3), (3, 1), (4, 3), and (5, 5). Since the domain of a function is the list of all possible x-values that it can have, the domain of this function is just { 1, 2, 3, 4, 5 }. While we're looking at this, let's also think about all the possible output values that this function can have—which is called the range of the function. Since this function is only defined at the five points shown on the graph, its range must be all the unique y-values that it can have. In other words, its range is { 1, 3, 5 }.

How about the function in the middle? Well, that function isn't just points on a plot, but lines too, so it's defined not only at the points in dark circles, but at all the points on the lines connecting the circles too. Which means that the domain of this function consists of all the points between 1 and 5. Or, using inequalities, we can say that the domain of this function is $1 \leq x \leq 5$. And its range? Well, since it has values anywhere along the line, its range is described by $1 \leq y \leq 5$.

The function on the right has arrows that indicate that it continues off in both direction beyond the edges of our plot. And as you can see, it is defined by the relationship $z(x) = (x^2 - 3) + 1$. Notice that no matter what value of x you plug into $z(x) = (x^2 - 3) + 1$, you'll always get an output number. Which means that the domain of the function is said to be "all real numbers." In other words, the domain is every number that you can find on the number line. How about its range? Well, as you can see by looking at the graph, the minimum value of the function is at the point (3, 1), and it continues to infinity in the positive z direction. Therefore its range is $z \geq 1$.

✎ POP QUIZ: WHAT'S THE DOMAIN AND RANGE?

Use your domain-and range-finding skills to figure out the domain and range of these four functions (they are all functions, right?).

Cont'd

Cont'd from page 286

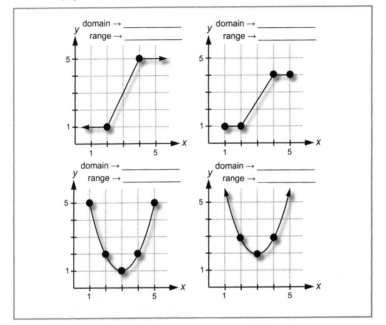

VISUALIZING POLYNOMIAL FUNCTIONS

Let's now put everything we've learned so far in this chapter together to see how we can picture what polynomial functions look like. In this section, we'll focus on graphing monomials of different degrees. In other words, we're going to make graphs of the simplest possible polynomials. Why? Because in the next section we'll see how we can use these simple plots, along with a bit of clever arithmetic, to graph *any* polynomial.

OVERACHIEVER  BADGE

FUNCTION VS. EQUATION

Are you bothered or confused by the use of the words "function" and "equation"? If so, you may want to read this . . . because it should help clear things up. We're about to plot lots of things that look like

$y=x$ and $y=x^2$

In the past, we've called things that look like this *equations*. But now I'm calling them polynomial *functions*. So which are they?

Well, the slightly inconvenient truth is that they could be either. How's that? When we think of $y=x^2$ as an equation, we're saying it's a thing that tells us that the expression on the left must have an identical value to the expression on the right. And that requirement limits the possible values that x and y can have for the equation to be true. On the other hand, we can also define the function $y(x)=x^2$. As we've discussed, a function like this doesn't tell us anything about the equality of the left and right sides. Instead, this function defines a process that turns input (called x) into output. In this case that output is equal to the square of the input.

But here's where things start to get a little confusing. We can take the output of our function, $y(x)$, and assign it to a variable. In other words, we can write an equation that says $y=x^2$ where we actually use the function $y(x)=x^2$ to calculate the right side of the equation. So, we can say that the equation $y=x^2$ is equivalent to the equation $y=y(x)$. Which means that while plotting the "equation" $y=x^2$ and the "function" $y=x^2$ are technically different things, you can now see that they end up producing the exact same result.

Cont'd

Cont'd from page 288

> Okay, now that we have that straightened out, let's get on with plotting some polynomial functions . . . or equations . . . or whatever.

WHAT DO CONSTANT FUNCTIONS LOOK LIKE ON A GRAPH?

Up first are degree 0 monomials. In other words, these are polynomials with only a single term, and where that term is a constant—which means there are no variables in these polynomials at all. In equation form, this type of polynomial looks like

$$y = C$$

"C" here represents whatever constant you want it to be—from as large a number as you'd like to as small a number as you can imagine. This constant can be an integer, a rational number, or something irrational . . . anything goes. Here's how to picture these constant functions graphically:

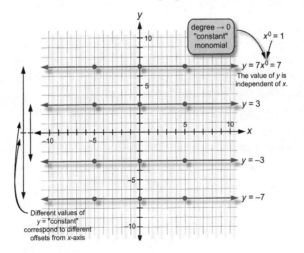

You can see four different functions plotted here. From top to bottom, they are

$y = 7$, $y = 3$, $y = -3$, and $y = -7$

As you can see, each of these functions is nothing more than a straight horizontal line parallel to the x-axis. And this makes perfect sense since there's no x in the function $y = C$. Which means that no matter where you look along the x-axis, the function always has the same constant value.

Q&DT ▶ The most important thing to notice here is that the different values of "C" shift the horizontal line up and down away from the x-axis by different amounts. In other words, the horizontal line representing the function $y = 3$ is always positioned 3 above the x-axis, the line for $y = -3$ is always positioned 3 below the x-axis, and so on.

WHAT DO LINEAR FUNCTIONS LOOK LIKE ON A GRAPH?

Okay, so that's what we get for degree 0 monomials. How about the next step up the complexity ladder: degree 1 monomials. In equation form, this type of polynomial can be written as

$y = mx$

So what does this mean? Well, it says that the value of y must be equal to some constant number, m, times the variable x. And, as we saw in chapter 3, this is just the equation for a straight line . . . which explains why degree 1 polynomials are called "linear." Here's what these functions look like when plotted

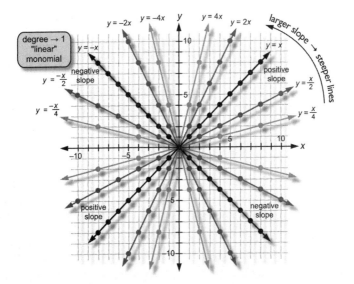

Wow! No, this isn't just one polynomial, there are actually ten different versions of the function $y = mx$ shown on this graph! What's the difference between them all? Believe it or not, the only difference is the value of the constant, m, which is called the "slope" of the line.

If you didn't know it already, slope is a really big deal in algebra. You've probably seen it used before when talking about the equation of a line—as in $y = mx + b$ (which we'll talk more about in a minute). The idea of slope comes up all the time in math, so you're likely to see it make appearances in class, on tests, maybe the SAT, and generally all over the place! So let's take a few minutes to dive in a bit deeper and talk about the details of slope.

WHAT DOES SLOPE MEAN?

When m > 0, the slope is positive, which means that the function $y = mx$ produces lines that run diagonally from upper right to lower left. You can see the particular values of m for each of the five lines with positive slopes shown on the graph. For example, the black line has a slope of 1, the lines above and below it have slopes of 2 and $\frac{1}{2}$, and the lines above and below those have slopes of 4 and $\frac{1}{4}$.

On the other hand, when m < 0, the slope is negative and the function $y = mx$ produces lines that run diagonally from upper left to lower right. As with positive slopes, there are five lines on this plot with negative slopes: the black line has a slope of –1, the lines above and below it have slopes of –2 and $-\frac{1}{2}$, and the lines above and below those have slopes of –4 and $-\frac{1}{4}$.

So what does the slope really mean? Well, the simplest way to think about the slope of a line is that it tells you how much a line rises or falls in the y direction when you move a distance of one in the positive direction of the x-axis. For example, the slope of the line $y = x$ is 1—which means that each time you move one step to the right along the x-axis, the line also moves one step up along the positive y-axis. Similarly, the line $y = 2x$ has a slope of 2, which means that every step along the x-axis results in a two-step rise along the positive y-axis.

Negative slopes can be interpreted in the exact same way, only this time moving one step along the positive x-axis produces a fall (instead of rise) along the y-axis by an amount equal to the slope. So for the line with slope equal to –4, moving one step along the positive x-axis produces a fall of four steps along the y-axis. The key thing to notice about this family of lines is that as the slope gets more and more positive or more and more negative, the lines become steeper and steeper.

Q&DT ▶

HOW TO CALCULATE SLOPE?

So that's what the slope of a line means, but how do you calculate it? In other words, how can you go from looking at a plot of a line like this

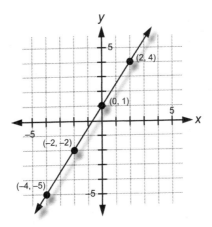

to actually figuring out what its slope is? Well, all we really need to do is go back to the meaning of slope. So far we've been saying that the slope tells you how much a line rises or falls when you move one step along the positive x-axis. But the slope of a line actually tells you a little more than that. If you take any two points on a line—we'll pick two points with coordinates (x_1, y_1) and (x_2, y_2)—then the slope of that line is given by

$$m = (y_2 - y_1) / (x_2 - x_1)$$

In other words, the slope of a line doesn't change as you go along the line, so not only can you calculate the slope by looking at the rise or fall over a distance of one step along the x-axis (which is just taking $x_2 - x_1 = 1$ in this equation), but you can calculate the slope by looking at the *total* rise or fall over *any* distance along the x-axis.

Sometimes you'll see the symbols Δx and Δy (pronounced "delta-x" and "delta-y") used to describe the "run" and "rise" differences along the x and y-directions. In other words, $\Delta x = x_2 - x_1$ and $\Delta y = y_2 - y_1$. Which means that you'll also sometimes see the slope of a line written as $m = \Delta y / \Delta x$ or refered to as "rise over run." So if you see that in your math book, that's what it means. ◀Q&DT

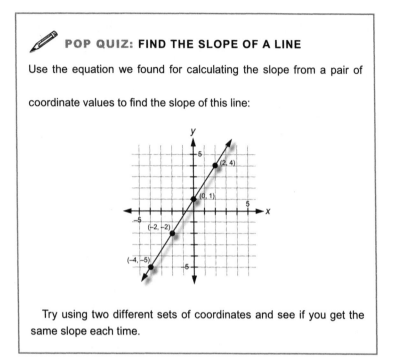

POP QUIZ: **FIND THE SLOPE OF A LINE**

Use the equation we found for calculating the slope from a pair of

coordinate values to find the slope of this line:

Try using two different sets of coordinates and see if you get the same slope each time.

WHAT DO QUADRATIC AND CUBIC FUNCTIONS LOOK LIKE ON A GRAPH?

We'll have more to say about linear functions and lines a bit later, but let's continue our march through the world of monomials and take a look at those with degrees of 2 and 3. In other words, let's take a look at what are called quadratics like

$$y = ax^2$$

and cubics like

$$y = bx^3$$

In these equations, the variable x is raised to either the second or third power and multiplied by a constant. Remember, the constant numerical coefficient of a line turned out to be its all-important slope. So what do the constants for quadratics and cubics mean? Well, before I tell you, take a look at these graphs and see if you can figure it out:

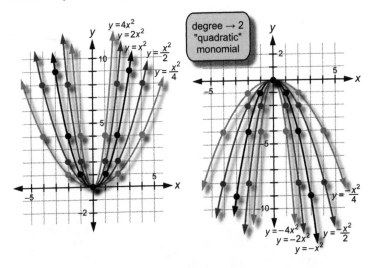

The graph on the left shows the behavior of quadratic mono-mials with positive constants, and the graph on the right shows the behavior of those with negative constants. All of the curves you see are called parabolas. And, as you can see, those that have positive constants open upward, while those that have negative constants open downward. What about the magnitude of the constant—what does that do? Well, if you look at the parabolas with larger numerical constants, you'll see that they rise more rapidly as you move in both the positive and negative x directions. In other words, the larger the magnitude of the numerical constant, the more "closed" the parabola.

How about cubics?

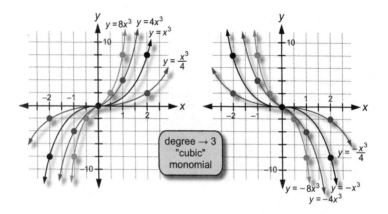

The family of cubic monomials shown on the left have positive numerical constants, and those on the right have negative constants. You can see that the larger the magnitude of the constant, the more quickly the curve rises or falls as you move along the *x*-axis away from the origin. So the larger the magnitude of the numerical constant, the more "squished" toward the *y*-axis the cubic function will be.

Okay, that's it for our run-through of the behavior of monomials. Of course there are higher degree monomials than cubics, but I'll leave it to you to explore their behavior in the Final Exam at the end of the chapter.

HOW TO SOLVE PROBLEMS WITH POLYNOMIALS

One great thing about polynomials is you can add, subtract, or multiply any two of them, and the result will always be another polynomial. And not only is that a handy feature, but it also means that we can use the simple monomials that we plotted earlier to build more and more complex polynomials.

HOW TO ADD POLYNOMIALS

Let's begin our look at polynomial arithmetic in the most logical place—polynomial addition. To get a feel for what it means to add polynomials, let's use the two monomials $y=2$ and $y=x$ to build the binomial $y=x+2$. How can we do it? Well, take a look at this:

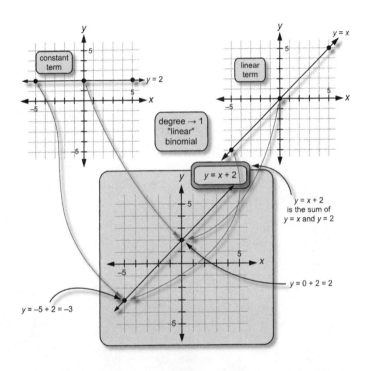

At the top, you can see the graphs of the two simple monomials $y=2$ and $y=x$ . . . exactly as we saw previously. Now here's the cool part: We can build the graph of the binomial $y=x+2$ from these two monomials by adding together the two graphs. As you can see in the plot at the bottom of the drawing, if we move point by point along the x-axis and add together the values of the two monomials at those points, we end up creating the plot of the binomial $y=x+2$ that's shown by the black line! Pretty cool, right?

HOW TO SUBTRACT POLYNOMIALS

Just as we made a graph of the binomial $y=x+2$ by adding together the monomials $y=x$ and $y=2$, we can do the same thing to create even more complex polynomials too. And we're not restricted to adding monomials to do so, we can subtract them as well. For example, let's think about how to graph the quadratic polynomial

$$y=x^2-x-4$$

How should you proceed? Well, throughout the book we've been resorting to creating a table at this point that gives a bunch of x-values and their corresponding y-values, and then we've produced a graph by plotting these points. But now we know that we can instead add or subtract several easy-to-plot monomials and get the same result much faster. To see how this works, let's first create graphs for each of the monomials, and then we'll add and subtract these graphs as needed to obtain the graph of the polynomial, like this:

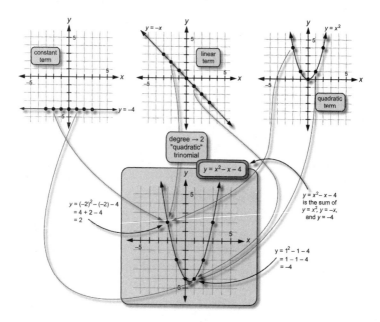

Take a minute to study how the three monomial terms combine together to create the quadratic trinomial. In particular, notice how the constant terms create an overall offset in the vertical direction, and the linear term shifts the parabola along the positive x-axis. Which direction would the parabola be shifted if the linear term were added rather than subtracted? You'll have your chance to answer that question in the Final Exam at the end of the chapter.

OVERACHIEVER ✪ BADGE

POLYNOMIAL MULTIPLICATION AND DIVISION

Just as we can add and subtract polynomials, we can also multiply and divide them. When we multiply a polynomial by a numerical constant, we simply scale each coefficient of the polynomial. For example, if we multiply the polynomial expression $x^2 - x - 4$ by the constant monomial 2, the resulting polynomial is

$$2 \bullet (x^2 - x - 4) = 2x^2 - 2x - 8$$

If you were to graph this new polynomial, you'd find that the new function looks exactly like the old one, except that it's stretched out by a factor of two in both the positive and negative y-direction (you should try it yourself). Of course, we can also multiply more complicated polynomials using the distributive property and FOIL, and we can even divide them.

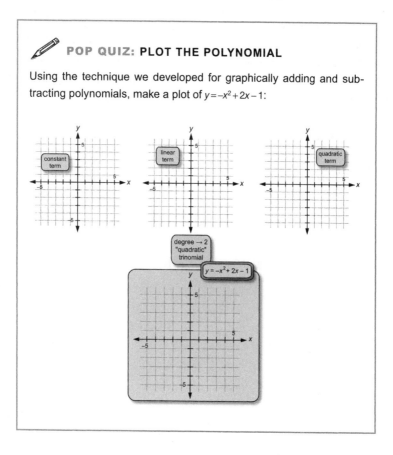

✏️ **POP QUIZ: PLOT THE POLYNOMIAL**

Using the technique we developed for graphically adding and subtracting polynomials, make a plot of $y = -x^2 + 2x - 1$:

• **ALGEBRA TUTORIAL** •

HOW TO FIND AND WRITE THE EQUATIONS OF LINES

Now that we've learned about polynomials, how to do arithmetic with them, and how to plot them, it's time for a more thorough discussion of what is arguably the most important type of polynomial. Which type am I referring to? Lines, of course. But why lines . . . what makes them so special? Well, as we saw in chapter 3, a lot of problems in the world can be solved with linear equations. And the basis of a linear equation is, of course, the equation for a line. So, let's learn how to find, write down, and work with the equations for lines. Up first, let's learn about the two main forms for writing equations of lines.

FORM #1: "SLOPE-INTERCEPT FORM"

All of the equations for lines that we looked at in chapter 3 were written like

$$y = mx + b$$

in what is known as "slope-intercept" form. As you can see, the expression on the right side of the equation is a degree 1 binomial. Which means that we can use what we've learned about adding polynomials to help us understand what the two constants, m and b, mean.

In particular, we found that plotting an equation like $y = C$ produces a horizontal line that's offset from the x-axis by an amount C. And we also found that plotting an equation like $y = mx$ produces a line (no surprise there!) with a slope, m, that determines the direction and steepness of the line. So, using the polynomial arithmetic skills we developed earlier, we can put these two things together (literally) and figure out that m and b must be the slope (which you're already familiar with) and what's called the "y-intercept" of the line.

Slope-Intercept Form

when $y = 0$
$$x = -\frac{b}{m}$$
x-intercept

$$y = mx + b$$
slope

when $x = 0$
$$y = b$$
y-intercept

So what's the y-intercept? Well, if you take a look at the plot of the line $y=x+2$ below, you'll see that the line crosses the y-axis at the point (0, 2). Which means that the y-intercept is just the location where the line crosses (or intercepts) the y-axis. Since the y-axis is located at the point $x=0$, when we plug $x=0$ into the equation $y=mx+b$, we get $y=b$ . . . in other words, it's the ordered pair (0, b).

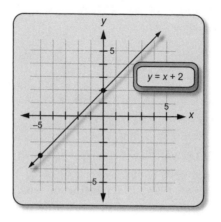

We can also find the location where the line crosses the x-axis by setting $y=0$ in the equation $y=mx+b$ (since the x-axis is located at $y=0$). When we do that, we get the ordered pair $(-b/m, 0)$, which is called the x-intercept.

FORM #2: "STANDARD FORM"

But slope-intercept form isn't the only way to write an equation for a line. For example, the equation

$$Ax+By=C$$

also represents a line. This way of representing a line is called "standard form," and you'll see it used quite often. In standard form, the three constants A, B, and C must all be integers (which means that no fractions are allowed), and traditionally A>0.

Now, the fact that the same line can be represented in either slope-intercept or standard form means that we should be able to convert from one form to the other. So, let's try to do exactly that by solving our standard form equation for the variable y, and then seeing if we can get something that resembling $y=mx+b$. If we subtract Ax from both sides of Ax+By=C, and then divide both sides by B, we get

$$y=(C-Ax)\,/\,B$$

Now, dividing through by B and switching the order of the two terms on the right, we get

$$y=-(A/B)x+C/B$$

Which, you'll notice, does indeed look exactly like the equation $y=mx+b$. Oh, really? Yes, they look exactly the same . . . so long as m=–(A/B) and b=C/B. And those relations are also the key things for you to remember when you're trying to interpret the meaning of the constants of a standard form linear equation.

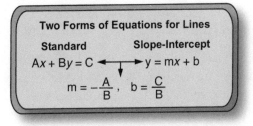

Two Forms of Equations for Lines

Standard   Slope-Intercept

$$Ax+By=C \longleftrightarrow y=mx+b$$

$$m=-\frac{A}{B}\,,\quad b=\frac{C}{B}$$

◄Q&DT

Now that we know about the two most common forms for writing the equation of a line (there are others, but they're less useful), let's begin to tackle some problems dealing with these equations.

PROBLEM TYPE #1: PUTTING EQUATION
INTO STANDARD FORM

The first problem that you may be confronted with is converting an equation into standard form. For example, let's say you're given the equation

$$2x + 8 = \left(\frac{5y}{2}\right) + x + 3$$

and asked to convert it into standard form. What's the best approach?

■ STEP 1: GET RID OF FRACTIONS AND DECIMALS

The first step in putting an equation into standard form is to get rid of any fractions or decimals from the equation. How do you do that? Well, if you have decimal numbers, first convert them into fractions. Then, successively multiply both sides of the equation by the denominators in the fractions. The right side of our problem has a fraction in the first term. So, let's multiply both sides by 2 to get rid of it. That gives us

$$4x + 16 = 5y + 2x + 6$$

■ STEP 2: MOVE ALL TERMS CONTAINING X AND Y TO THE SAME SIDE OF EQUATION

Since we ultimately want to put the equation into the form $Ax + By = C$, we now need to move all the terms containing either x or y to one side of the equation. So, let's subtract $4x$ from both sides of the equation to get

$$16 = 5y - 2x + 6$$

■ STEP 3: MOVE ALL CONSTANT TERMS TO THE OTHER SIDE OF THE EQUATION

Now we need to move all the constant terms to the other side of the equation. So, let's subtract 6 from both sides to get

$$10 = 5y - 2x$$

■ **STEP 4: MULTIPLY BOTH SIDES BY –1 IF NECESSARY TO MAKE A > 0**

Let's now turn the equation around so the variables are on the left, and let's put it into the form $Ax + By = C$:

$$-2x + 5y = 10$$

Are we done? Well, almost. The final step is to check and see if $A > 0$. If it's not, then we need to multiply both sides of the equation by –1 to make sure it is. In our case, $A = -2$, so we need to multiply both sides by –1. After doing that, we get our final equation for a line in standard form:

$$2x - 5y = 10$$

Here's a summary of the steps to use to put an equation in standard form:

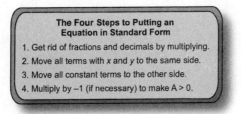

The Four Steps to Putting an Equation in Standard Form
1. Get rid of fractions and decimals by multiplying.
2. Move all terms with x and y to the same side.
3. Move all constant terms to the other side.
4. Multiply by –1 (if necessary) to make A > 0.

◀ Q&DT

PROBLEM TYPE #2: FROM PLOT TO EQUATION

A second type of problem dealing with the equations of lines that you may see is to create an equation for a line based on the graph of that line. In this part of the tutorial, we're going to come up with an equation for the following line:

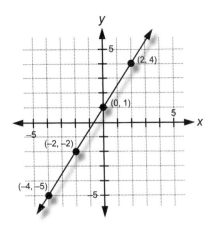

■ STEP 1: CALCULATE THE SLOPE OF THE LINE

The first step is to calculate the slope of the line. As we learned earlier, we can calculate the slope of a line by first finding the coordinates of two points on the line, (x_1, y_1) and (x_2, y_2), and then calculating

$$m = (y_2 - y_1) / (x_2 - x_1)$$

For our example, let's take the points (0, 1) and (2, 4) to calculate that the slope of the line is

$$m = (4 - 1) / (2 - 0) = 3/2$$

And it's a good thing we found $m = 3/2$, because that's the same answer we found for the exact same line earlier!

■ STEP 2: USE THE SLOPE AND ONE POINT TO FIND THE Y-INTERCEPT

The second step is to use this slope along with the coordinates of one point to come up with an equation. How does this work? Well, it's actually done using the exact same technique that we used to calculate the slope . . . with one small difference: Instead of needing the coordinates of two points, this time we only need the coordinates of one point. Start by writing an equation for the slope:

$$m = (y - y_1) \, / \, (x - x_1)$$

This says that we can choose any point on the line with coordinates (x, y) and plug its values into this equation to get the slope. But we've actually already calculated the slope, so we don't need the equation for that. Instead, let's multiply both sides of this equation by $(x - x_1)$ and flip it around to get

$$y - y_1 = m \bullet (x - x_1)$$

Then, let's multiply out the right side and move the y_1 from the left to the right to get

$$y = y_1 + mx - mx_1$$

And finally, let's rearrange the terms to put this equation into slope-intercept form, like this:

$$y = mx + (y_1 - mx_1)$$

By comparing this to $y = mx + b$, you can see that we can now write an equation for the y-intercept

$$b = y_1 - mx_1$$

Okay, so what have we done? Well, we've just managed to figure out a way to take the slope of a line and the coordinates of one point on that line, and use some algebra to turn those into m and b—which we can use to write an equation for the line in slope-intercept form. For our problem, we've already calculated that m = 3/2, so if we now use the coordinates of the point (2, 4) for the values of (x_1, y_1), we can calculate b like this:

$$b = 4 - (3/2 \bullet 2) = 4 - 3 = 1$$

Of course, if we can see what the y-intercept is just by looking at the plot, then we don't have to do this step at all. But you can't always rely on being able to do that since the y-intercept often does not fall at an integer value of y. So you need to know how to use this more general technique.

■ STEP 3: CONSTRUCT YOUR FINAL EQUATION IN SLOPE-INTERCEPT FORM

The last step is the easy part—all you have to do is construct your final equation in slope-intercept form using the values of m and b that you've found in the two previous steps. In our case, the final equation is

$$y = \frac{3}{2}x + 1$$

■ STEP 4: CONVERT TO STANDARD FORM IF NECESSARY

If you're asked for the equation in standard form instead of slope-intercept form, you would now just go through the procedure we outlined in the earlier part of this tutorial to put an equation into standard form.

To sum up, here's a summary of the steps required to create an equation for a line from its graph:

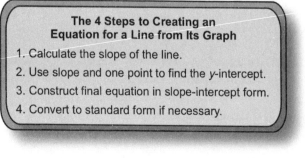

The 4 Steps to Creating an Equation for a Line from Its Graph

1. Calculate the slope of the line.
2. Use slope and one point to find the y-intercept.
3. Construct final equation in slope-intercept form.
4. Convert to standard form if necessary.

Q&DT ▶

There are many variations of these problems related to finding the equations of lines. For example, instead of being presented with the graph of a line and being asked to come up with an equation for it, you could instead be given the slope of a line and the coordinates of a point on it, or just the coordinates of two points on the line, and then asked to find an equation. But, if you think about it, you'll see that each of these cases requires you to do the same steps that we just went through. So if you learn the techniques we've covered, you'll be in good shape for whatever else comes your way.

EQUATIONS OF HORIZONTAL AND VERTICAL LINES

Before finishing up our discussion of the equations of lines, let's quickly talk about a few special cases. Up first, horizontal lines. If you look at the slope-intercept form of the equation for a line, $y = mx + b$, you'll see that when $m = 0$ the equation just becomes $y = b$. In other words, it's exactly like the equations of constant monomials that we looked at earlier. Which means that you already know how to work with and write the equations of horizontal lines!

The second special case in the world of lines is a line that's parallel to the y-axis—in other words, a vertical line. Why is this special? For one thing, this is the only type of line whose equation cannot be written in the form $y = mx + b$. Why not? Well, try and figure out what the slope of a vertical line is. We know that the slope of a line is the amount that it rises or falls as you take one step in the positive x-direction. But this idea doesn't make sense when we're talking about a vertical line. So what's the answer then? Well, if the vertical line in question goes through the point $(2, 0)$, then the equation that describes that line is $x = 2$. This says that no matter what the value of y is, x is always 2.

So, to sum up these two special cases, we can say that:

- **Horizontal lines** parallel to the x-axis have zero slope and are described with equations like $y =$ "constant"
- **Vertical lines** parallel to the y-axis do not have a slope (the slope is undefined) and are described with equations like $x =$ "constant"

POP QUIZ:
FIND THE EQUATION OF A LINE . . .
THEN PLOT IT!

Use the techniques outlined in the last tutorial to find the equation for the line (in slope-intercept form) that has a slope of $\frac{1}{2}$ and passes through the point (3, 2).

Now plot this line using the slope and y-intercept that you find:

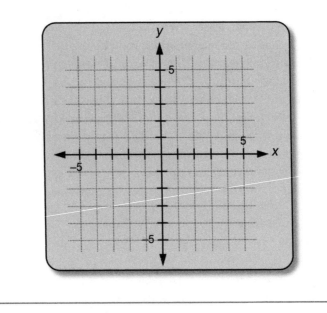

SECRET AGENT MATH-LIB

WHAT'S THAT DUDE SAYING?

Good afternoon, secret agent _____ *(name of fruit or vegetable)*. Is the service a little slow today? You've been waiting thirty minutes and still no pasta! Just as you're about to pull out a(n) _____ *(name of fruit or vegetable)* to snack on, the waiter drops off your lunch…and a note. A note? That's a little odd. But when you look at the note, you see that it's a lot more than a little odd…it's bizarre! The message says:

G O   F I G H T   W I N !

You look at the waiter with a puzzled expression, but he points to a man at the back of the diner whom you immediately recognize to be someone you work with. What's his name? Oh yeah. It's secret agent _____ *(name of fruit or vegetable)*. Needless to say, you're now more confused, and a little worried to boot since you're sure this isn't just a cheerful and encouraging note. In fact, you're sure that it's a secret message.

Without hesitating you pull out your code-cracking handbook, turn to the first page, and see the following prompts and puzzles, as well as the plot on the following page:

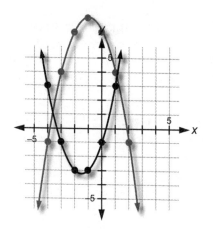

Your job is to answer the following questions...write your answers on the first blank line (we'll come back to the second blank line in a minute):

Are the following expressions even (E) or odd (O) if x is an odd number?

1. $x^2 + 3x - 1$ .... <u> O </u> → <u> O </u>

2. $-x^2 - 2x + 7$ .... <u>   </u> → <u>   </u>

Are the following expressions even (E) or odd (O) if x is an even number?

3. $x^2 + 3x - 1$ .... <u>   </u> → <u>   </u>

4. $-x^2 - 2x + 7$ .... <u>   </u> → <u>   </u>

Are the following expressions positive (P) or negative (N) if $x > 2$?

5. $x^2 + 3x - 1$ .... <u> P </u> → <u> O </u>

6. $-x^2 - 2x + 7$ .... <u>   </u> → <u>   </u>

Are the following expressions positive (P) or negative (N) if $-1 < x < 0$?

7. $x^2 + 3x - 1$ .... _____ → _____

8. $-x^2 - 2x + 7$ .... _____ → _____

Which property allows you to make the following changes without changing the values that the polynomial evaluates to. Is it the commutative property of addition (A) or the commutative property of multiplication (M)?

9. $x^2 + 3x - 1$ to $3x + x^2 - 1$ .... _____ → _____

10. $-x^2 - 2 \cdot x + 7$ to $-x^2 - x \cdot 2 + 7$ .... _____ → _____

Excellent job! Now, all you have to do is use your answers to these ten problems to crack the code. Of course, in classic secret agent math fashion, that's not as straightforward as you might hope. So, what to do?

The first step is to fill in the final blank spaces next to the letters on the right side of each problem. For every letter that occurs more than once, write a zero in the blank space. That should leave three problems with unfilled blank spaces.

- In the first remaining blank space, write the number you get by multiplying together the numbers of the problems that gave you the first "E" and the first "N".
- In the second remaining blank space, write the number you get by adding together the numbers of the problems that gave you the last two "O"s.
- In the last remaining blank space, write the number you get by adding 1 to twice the number of the problem that gave you the letter "M".

And with that, you're now ready to decrypt that message! For each letter in the encrypted message "GO FIGHT WIN," cycle through the alphabet the number of letters indicated by the answer to the problem shown in parentheses beneath the letter. In other words, for the second letter in "GO FIGHT WIN," "O", we need to cycle through the alphabet zero letters (since the number in the last blank space of problem five is zero). Continue to cycle through the alphabet for each letter to decrypt the message.

So, what's that dude saying?

G　O　　F　I　G　H　T　　W　I　N!

$\underline{\hspace{1em}}$ $\underline{\overset{O}{\hspace{1em}}}$　$\underline{\hspace{1em}}$ $\underline{\hspace{1em}}$ $\underline{\hspace{1em}}$ $\underline{\hspace{1em}}$ $\underline{\hspace{1em}}$ ．$\underline{\hspace{1em}}$ $\underline{\hspace{1em}}$ $\underline{\hspace{1em}}$!
(9) (5)　(1) (8) (3) (7) (4)　(10) (2) (6)

| | | | M | A | T | H | | B | R | A | I | N | | G | A | M | E | | | |

THE MARCH OF THE MONKEYS

Few things in life are guaranteed, but I can pretty much assure you that the following will never happen to you. Imagine that you're trekking through the jungles of Costa Rica when you come across a boisterous group of howler monkeys all dutifully marching in a perfect line out in the distance in front of you. Next to you on the ground, you see a pile of bananas and a sign that politely asks you to toss them to the monkeys . . . but only in the precise direction indicated by an arrow on the ground. See, pretty strange scenario, right?

THIS PROBLEM IS BANANAS

The note next to the bananas says that the monkeys are walking along a line that satisfies the equation $2y - x = 8$, and that the direction you'll be throwing the bananas is along a line that satisfies $2x + y = 16$. And apparently, you've just realized, you're standing at the position marked (8, 0) on this coordinate system:

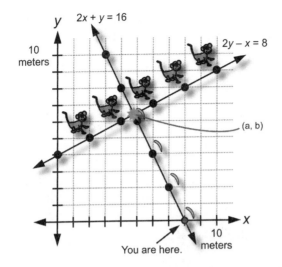

After seeing the setup, you can't resist tossing some bananas to the howlers. And while mindlessly tossing bananas, you begin to wonder how far away you're standing from the point at which the monkeys are intercepting your passes. Are you safe? How can you figure that out? Think about it, stare at the graph for a while, and then read on to find the answer.

➤ STEP 1: FIND THE X-VALUE OF THE INTERSECTION

Let's start by labeling the coordinate where the monkeys are catching the bananas—we'll call it (a, b). But how can you find out where that spot is on the big coordinate system? Well, what major event happens right at the spot where the monkeys catch the bananas? It's right where the two lines describing the trajectories of the monkeys and the bananas intersect! In other words, the coordinate (a, b) is the point at which both lines have the exact same x and y values.

Which means that in order to figure out how far away they are, we need to first find the special value of x at which both lines have the same value of y. How can we find it? Well, the first thing we should do is solve each of our two equations for y. In other words, let's write them in slope-intercept form.

The equation describing the monkey-marching line is $2y - x = 8$. To put this into slope-intercept form, all we have to do is add x to both sides and divide by 2 to get

$$y_{monkey} = \frac{1}{2}x + 4$$

Notice that I've put the subscript "monkey" on the variable for y here to remind us that this is the equation for the line followed by the monkeys. Next we have the equation describing the banana tossing line: $2x + y = 16$. We can put this in slope-intercept form simply by subtracting $2x$ from both sides to get

$$y_{bananas} = -2x + 16$$

As with the previous equation, I've added a subscript to this variable for y to remind us that this is the equation for the line describing the trajectory of the bananas.

Now that we have these two equations, we're ready to figure out what x value makes them both have the same y value. In other words, the problem we're trying to solve is

$$y_{monkey} = y_{bananas}$$

which means that we just need to set our two equations equal to each other:

$$\frac{1}{2}x + 4 = -2x + 16$$

The value of x that solves this equation is also the value that makes $y_{monkey} = y_{bananas}$. By this point you should be an expert at solving equations like this, so I won't go through all the details. Once you shuffle some constants and variables around, you'll find that the solution is $x = 24/5$. So we now know half of the ordered pair describing the point at which monkey meets banana: $(24/5, b)$. How about the other half?

➤ STEP 2: FIND THE Y-VALUE OF THE INTERSECTION

Once you've solved for the value of x at which the two lines intersect, all we have to do to find the y value of this point is to plug $x=24/5$ into either of the two equations—it doesn't matter which because they both have the same value at that point. So, if we plug $x=24/5$ into the equation for the monkey marching line, we get

$$b = \frac{1}{2} \cdot \left(\frac{24}{5}\right) + 4 = \frac{12}{5} + 4 = \frac{32}{5}$$

Now, just to make sure we've got the right ordered pair, you should try plugging $x=24/5$ into the other equation too—the one for the banana trajectory line. The value for b that you get should be the same. Which means that the point where monkey and banana collide is described by the coordinate (24/5, 32/5).

➤ STEP 3: FIND THE DISTANCE TO THE MONKEYS

We now know the x and y values where the monkeys catch the bananas, so all that's left to do is figure out how far away the point $(a, b) = (24/5, 32/5)$ is from the point at which you're standing, $(8, 0)$. And all we need to do to find that distance is to use our old pal the Pythagorean theorem.

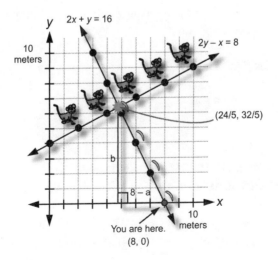

Take a look at this drawing and you'll see a right triangle with a height equal to b and a width equal to $8 - a$. The length of the hypotenuse of that triangle is precisely the number we're after—it's the distance between you and the place where the monkeys are catching the bananas. Using the Pythagorean theorem, we can find that this distance is given by

$$\text{distance} = \sqrt{b^2 + (8 - a)^2}$$

Let's now plug in the values of the coordinate (a, b) to find the actual distance:

$$\text{distance} = \sqrt{\left(\frac{32}{5}\right)^2 + \left(8 - \frac{24}{5}\right)^2}$$

Using the fact that $8 = 40/5$ gives us

$$\text{distance} = \sqrt{\left(\frac{32}{5}\right)^2 + \left(\frac{40}{5} - \frac{24}{5}\right)^2} = \sqrt{\left(\frac{32}{5}\right)^2 + \left(\frac{16}{5}\right)^2}$$

Squaring everything and then doing the addition, we get

$$\text{distance} = \sqrt{\frac{1280}{25}} = \sqrt{\frac{256}{5}}$$

Since the distances along the axes is given in units of meters, this gives us a distance of about

$$\text{distance} \approx 7.16 \text{ meters}$$

which is about 23.5 feet. We should be safe!

THE SLOPES OF PARALLEL AND PERPENDICULAR LINES

Since we spent a good part of this chapter talking about slopes, it seems appropriate to mention one final slope-related piece of information that made an appearance in the math brain game. What was it? Well, did you notice that the two lines in this problem intersected each other at a right angle? When that happens, we say that the two lines are "perpendicular" to each other. So what does that have to do with slopes?

Well, you can tell if two lines are perpendicular to each other just by looking at their slopes. If the slope of the first line, which we'll call m_1, and the slope of the second line, which we'll call m_2, happen to satisfy the equation $m_2 = -1 / m_1$, then those two lines are perpendicular. This type of relationship, where one number is the negative inverse of another, is often called the "negative reciprocal." If you take a look at the slopes of the lines in our monkey problem ($\frac{1}{2}$ and -2), you'll see that these two numbers are indeed negative reciprocals of each other, and are therefore the slopes of perpendicular lines.

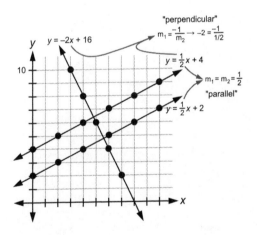

On the other hand, two lines that never cross each other are called "parallel" lines. What's the trick to tell if two lines are parallel? It's even easier than the trick for checking to see if they're perpendicular. Parallel lines are any lines that have the exact same slope. So, for two lines with slopes m_1 and m_2, if $m_1 = m_2$ then those two lines are parallel. To sum up:

Q&DT ▸

- Two lines that are **perpendicular** cross at right angles and have slopes that satisfy the equation $m_1 = -1 / m_2$. In other words, the two slopes are the negative reciprocals of one another. For example: the slopes 2 and $-1/2$.
- Two lines that are **parallel** never cross and have slopes that satisfy the equation $m_1 = m_2$. In other words, they have the exact same slope.

SYSTEMS OF EQUATIONS

While the scenario put forward in the monkey and banana problem was admittedly a little silly, the algebra in it was not silly at all. In fact, that problem contained our first look at doing something called solving a system of equations. The word "system" here just refers to the fact that we're solving more than one equation at a time.

More specifically, the problem we looked at is an example of what's called a system of *linear* equations—since both equations in the problem were linear. But a system of equations doesn't have to contain only linear equations. You can have systems of quadratic equations, or cubic equations, or even systems of equations with a bunch of different degrees. The degree of the various equations in the system doesn't matter . . . but the number of equations in your system most certainly does.

NUMBER OF EQUATIONS = NUMBER OF UNKNOWNS

So how many equations can you have in a system? Well, there's actually no maximum. But there is a minimum. The quick and

Q&DT ▸

dirty tip is that in any system of equations, you need to have as many equations relating the unknown variables in the problem as you have unknown variables. For example, in a problem with

- one unknown, you need one equation
- two unknowns, you need two equations

- three unknowns, you need three equations
- four unknowns, you need four equations

and so on. Which explains why we never needed more than one equation to get an answer to the single variable equations we solved back in chapter 2. And it also explains why we needed two equations to solve the monkey and bananas problem—namely because there were two unknown positions in that problem: a and b. But enough with the introduction, let's move on to looking at how to solve systems of equations.

• ALGEBRA TUTORIAL •

HOW TO SOLVE A SYSTEM OF EQUATIONS

There are three different methods that you can use to solve a system of equations. All three methods will work to solve any problem, but some methods definitely are better suited to certain problems than others.

METHOD ONE: GRAPHICAL SOLUTION

The first method we'll discuss is to solve a system of equations graphically. While this method is excellent for helping you visualize and better understand what you're doing (which is why we've used variations of it throughout the book), it is also much slower and more cumbersome than the other two methods we'll talk about. Plus, it only works when you're dealing with equations that you know how to graph! Which also means that it doesn't work well when there are more than two unknowns—since it's hard to visualize plots in more than two dimensions.

In this section of the tutorial, we'll demonstrate the process of solving a system of equations graphically using the pair of equations

$2y = x^2 - 2$ and $4y = 2x$

■ **STEP 1: PUT EACH EQUATION INTO A FORM WHICH YOU CAN GRAPH**

The first step when solving a system of equations graphically is to put all of the equations in a form that you can easily graph. In this case, we need to solve each equation for y, like this:

$$y = \frac{1}{2}x^2 - 1 \text{ and } y = \frac{1}{2}x$$

■ **STEP 2: GRAPH EACH EQUATION ON THE SAME SET OF AXES**

The second step is to actually graph all of the equations. Make sure you cover a wide-enough region of the x and y-axes so that you can see all the locations where the two curves intersect (since those points will be the solutions to our system of equations). Here's the graph for our problem:

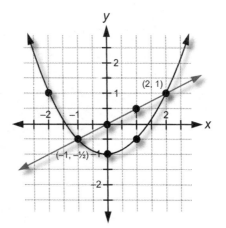

■ **STEP 3: FIND THE LOCATION(S) WHERE THE LINES INTERSECT**

As you can see by looking closely at our plot, the two curves intersect at the points (–1, –1/2) and (2, 1). But what if the points hadn't fallen at these nice neat values that allowed us to tell exactly where the intersections are located? Well, that's one of the complexities of trying to solve systems of equations graphically that you have to deal with. Most of the time, the best way to deal with it is to move on and try solving the problem using one of the other solution methods. But before we do that, here's a summary of the three-step process for solving a system of equation graphically:

◄ Q&DT

> **The Three Steps to Solving a System of Equations Graphically**
> 1. Put each equation into a form which you can easily graph.
> 2. Graph each equation on the same set of axes.
> 3. Find the location(s) where the lines intersect.

METHOD TWO: SOLVING BY SUBSTITUTION

Next up on the list is a method called solving by substitution. This method is most useful when you're solving a system of linear equations for two variables. In other words, when you have two linear equations and two unknowns. To demonstrate the method, let's look at the system of equations

1. $3x - y = -1$
2. $2x + 2y = 8$

■ **STEP 1: SOLVE ONE EQUATION FOR ONE VARIABLE**

The first step in this method is to solve one of the equations for one of the variables. It doesn't matter which . . . in the end, you'll get the same answer no matter how you do it. But some choices will make your life easier since they'll require less work. In this case, I'm going to choose to solve the first equation for the variable y, since it's practically done for me already:

$$y = 3x + 1$$

■ STEP 2: SUBSTITUTE THE EXPRESSION FOR THE FIRST VARIABLE INTO THE SECOND EQUATION

The second step is to now take this solution for the variable y in the first equation, and substitute it into the second equation. In other words, we're going to plug $y = 3x + 1$ into $2x + 2y = 8$, like this:

$$2x + 2 \bullet (3x + 1) = 8$$

■ STEP 3: SOLVE THE SECOND EQUATION FOR THE VALUE OF THE SECOND VARIABLE

Now, in the third step let's solve for the variable x. First we need to multiply out the left side to get

$$2x + 6x + 2 = 8$$

And now we can simplify and solve for x

$$x = \frac{6}{8} = \frac{3}{4}$$

■ STEP 4: SUBSTITUTE THIS SOLUTION INTO THE FIRST EQUATION TO SOLVE FOR THE VALUE OF THE OTHER VARIABLE

We now know half of the solution. In other words, we know the value of one of the two variables. But we still need to figure out what y is equal to. To find out, we now need to take the numerical value of *x* and plug it back into the first equation. Doing so gives us

$$3 \bullet 3/4 - y = -1$$

If we simplify this and solve for *y*, we get

$$y = 1 + 9/4 = 13/4$$

So the solution to our problem is

$$x = 3/4 \text{ and } y = 13/4$$

■ STEP 5: CHECK YOUR SOLUTION!

Before moving on, you should try plugging these values into both equations and checking to make sure that the solutions work. Here's a summary of the five-step method for solving a system of equations by substitution:

> **The 5 Steps for Solving a System of Equations by Substitution**
> 1. Solve equation 1 for variable 1.
> 2. Substitute variable 1 into equation 2.
> 3. Solve equation 2 for value of variable 2.
> 4. Substitute value of variable 2 into equation 1 and solve for value of variable 1.
> 5. Check your solution!

◀Q&DT

METHOD THREE: SOLVING BY ELIMINATION

The last of our three methods for solving a system of equations is called solving by elimination. This method can actually be used to solve systems with more than two variables and more than two equations, although any more than three of each and things start to get difficult to do by hand. For this example, let's solve the exact same system of equations that we solved in the demonstration of the previous method:

1. $3x - y = -1$
2. $2x + 2y = 8$

■ STEP 1: MULTIPLY ONE OR BOTH EQUATIONS BY CONSTANTS

The goal with this method is to add or subtract the equations and get one of the variables to cancel out. But if we add or subtract the pair of equations right now, we won't get any variables to cancel. Which means that we first need to multiply both sides of one of the equations by a constant value so that a variable will cancel when we add them. In this case, let's multiply both sides of the first equation by 2 to get

$$6x - 2y = -2$$

■ STEP 2: ADD OR SUBTRACT THE EQUATIONS TO ELIMINATE A VARIABLE

We're now ready to add the two equations in order to eliminate a variable.

$$\begin{aligned} 6x - 2y &= -2 \\ 2x + 2y &= 8 \\ \hline 8x \phantom{{}- 2y} &= 6 \end{aligned}$$

As you can see, when we add the equations, the variable y cancels out, leaving us with the equation $8x = 6$.

■ STEP 3: SOLVE FOR THE VARIABLE THAT YOU DIDN'T ELIMINATE

The next step is to solve this new equation to obtain the value of one of the variables. In this case, the solution to $8x=6$ is just $x = \frac{6}{8}$. Which, equivalently, is equal to $x = \frac{3}{4}$.

■ STEP 4: PLUG THIS VARIABLE INTO EITHER EQUATION TO SOLVE FOR THE OTHER VARIABLE

We now have half the solution, and to get the other half all we have to do is plug $x = \frac{3}{4}$ into one of the two equations. In this case, let's plug $x = \frac{3}{4}$ into the first equation, $3x-y=-1$, to get

$$3 \bullet \frac{3}{4} - y = -1$$

If we solve this equation for y, we get $y = 1 + \frac{9}{4} = \frac{13}{4}$. So, once again, the solution to this system of equations is $x = \frac{3}{4}$ and $y = \frac{13}{4}$.

■ STEP 5: CHECK YOUR SOLUTION!

Before moving on, you should try plugging these values into both equations and checking to make sure the solution works. Here's a summary of the five-step method for solving a system of equations by elimination:

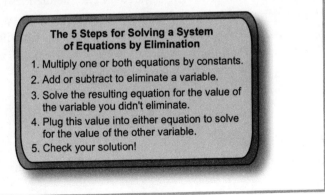

The 5 Steps for Solving a System of Equations by Elimination

1. Multiply one or both equations by constants.
2. Add or subtract to eliminate a variable.
3. Solve the resulting equation for the value of the variable you didn't eliminate.
4. Plug this value into either equation to solve for the value of the other variable.
5. Check your solution!

◀ Q&DT

POP QUIZ:

LET'S SOLVE A SYSTEM OF EQUATIONS!

Use the method of solving by elimination to find the solution to this system of three linear equations with three unknowns:

$2x + 3y + z = 1$ $\qquad -x - y + 2z = 2$ $\qquad x + 2y - 2z = 3$

SYSTEMS OF INEQUALITIES

Not to be left out of the fun, not only can we solve systems of equations, but we can also solve systems of inequalities. As you'll recall from chapter 3, inequalities are just like equations, except that instead of an equals sign they contain an inequality symbol: $>$, $<$, $\geq$, or $\leq$.

Whereas the goal of solving a system of two equations is to figure out the value of each of the two unknown variables in the problem, the goal of solving a system of two inequalities is to figure out which portion of the x-y plane contains (x, y) pairs that satisfy the combined requirements of the system of inequalities. The easiest way to see how this all works is with an example. So let's take a look at how to solve a system of inequalities.

HOW TO SOLVE A SYSTEM OF INEQUALITIES

The ultimate goal of this tutorial is to learn how to solve a system of inequalities. Throughout the tutorial, we're going to focus on the pair of inequalities

$y < x + 1$ and $y \geq -x + 3$

But before we begin working with them together, let's begin by taking a look at what they mean individually.

HOW TO GRAPH A LINEAR INEQUALITY

■ **STEP 1: PREPARE TO MAKE THE PLOT AS YOU WOULD FOR AN EQUALITY**

The first step in plotting a linear inequality has nothing to do whatsoever with inequalities. And that's because the first step is simply to prepare to plot the inequality as if you were actually plotting an equality. So, on the left I've prepared some points as a guide for plotting the line $y = x + 1$, and on the right I've prepared points for the line $y = -x + 3$.

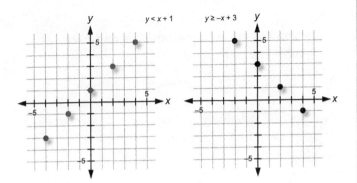

■ **STEP 2: PLOT THE LINE (SOLID IF INCLUDED OR DASHED IF EXCLUDED)**

The second step is to actually plot the line . . . exactly as you'd do when plotting an equality. But there's one key difference: If the line is to be included in the solution, then it should be drawn as a solid line; if it is not to be included in the solution, then it should be drawn as a dashed line to show this. How do you know if it should be included or not? Well, if the symbol in the inequality is either > or <, then the line is *not* included (since only values strictly less than or greater than the line are included). But if the symbol in the inequality is either ≥ or ≤, then the line should be included. Accordingly, the line on the left below is dashed, and the one on the right is solid:

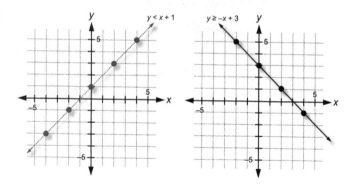

■ **STEP 3: SHADE THE APPROPRIATE REGION ABOVE OR BELOW THE LINE**

The final step in plotting an inequality is to shade the region either above or below the line to show which part of the coordinate plane is included in the solution. For example, for the inequality $y < x + 1$ on the left, the region below the line (not including the line) is part of the solution since those are all points with coordinates that satisfy $y < x + 1$. In other words, you can pick any point in that shaded region, plug it in to the inequality, and get a true statement. For example, if we plug the coordinate (4, 3) in, we get $3 < 4 + 1 \rightarrow 3 < 5$ ... which is certainly true. So, here's the finished product:

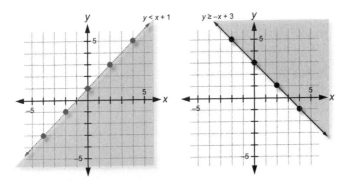

And here's a summary of the three-step method for creating a graph of a linear inequality

> **The 3 Steps for Creating a Graph of a Linear Inequality**
>
> 1. Prepare to plot the line just as you would for an equality.
>
> 2. Plot the line (solid if the line is included; dashed if it is not).
>
> 3. Shade the appropriate region above or below the line.

HOW TO COMBINE TWO INEQUALITIES TO MAKE A SYSTEM OF INEQUALITIES

By learning to plot these two inequalities, we've now done 90 percent of the hard work. The only thing left to figure out is how to go from dealing with two separate inequalities, to dealing with a system of inequalities. And to do that, you simply need to plot both inequalities in the system on the same set of axes using the three-step process we just went through. After doing that, the portion of the graph that is shaded by *both* parts of the system of inequalities is the region of the plane that satisfies the system—in other words, that doubly shaded region is your answer.

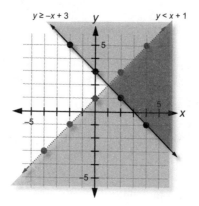

In our problem, the solution to the system of inequalities corresponds to the area on the right side of the drawing that is shaded both above the line $y \geq -x + 3$ and below the line $y < x + 1$. Every point in this doubly shaded region has coordinates that will simultaneously satisfy both inequalities. Try plugging in some points for yourself to make sure this is true!

POP QUIZ:

LET'S SOLVE A SYSTEM OF INEQUALITIES!

Use the following set of axes to find the area of the x-y plane that solves each of the following systems of inequalities:

1. $y > -x + 3$, $y < x + 1$, and $x \leq 4$
2. $y \geq -x + 3$, $y \leq x + 1$, and $y > 2$
3. $y \geq -x + 3$, $y < x + 1$, and $x < -2$

Cont'd

Cont'd from page 332

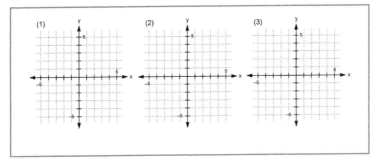

CHALLENGE PROBLEM #1

> *Pick a whole number greater than 2. Now multiply it by 2, and then subtract it from the square of your original number. Then add 1 to this result. After that, take the square root of this number, and then add 1 to the result. What's the answer? After all that work, did you get back exactly the same number you started with?*

Does that look familiar? It's probably been a while since you've read it, but if you look way back at the end of chapter 1 you'll see that it's the first "challenge problem" that I told you we'd be able to solve by the time we finished the book. Well, we're just about at the end of the book . . . which means that we'd better start solving them!

And, just in time, we now know enough algebra to understand how this riddle works. Here's what we're going to do: Let's talk through the sequence of steps in the riddle, and at the same time, let's write down an equation that mirrors those steps.

Riddle	Algebra	Example
"Pick a whole number greater than 2."	Create a variable called x	**5**
"Now multiply it by 2, and then subtract it from the square of your original number."	$x^2 - 2x$	$5^2 - 2 \cdot 5 = 25 - 10 = \mathbf{15}$
"Then add 1 to this result."	$x^2 - 2x + 1$	$15 + 1 = \mathbf{16}$
"Take the square root of this number."	$[\, x^2 - 2x + 1\,]^{1/2}$	$[16]^{1/2} = 4$
"Then add 1 to the result"	$1 + [\, x^2 - 2x + 1\,]^{1/2}$	$1 + 4 = \mathbf{5}$

As you can now see, by the time we reach the end of the riddle, we've created the rather complex-looking algebraic equation

$$1 + [x^2 - 2x + 1]^{1/2}$$

But that's not all we've done. If you look at the example sequence of numbers on the right, you'll see that the thing we started with, x, is equal to the thing we ended with, $1 + [\, x^2 - 2x + 1\,]^{1/2}$. Which means that

$$x = 1 + [x^2 - 2x + 1]^{1/2}$$

But other than the fact that the numbers end up matching, how can we know for sure that these are equal? The first way is to "undo" the equation by first subtracting –1 from both sides to get

$$x - 1 = [x^2 - 2x + 1]^{1/2}$$

and then squaring both sides to get

$$(x-1)^2 = x^2 - 2x + 1$$

All you have to do now to see that these are equal is to use FOIL to multiply out the expression on the left . . . which I'll leave for you to do.

So now you know the secret. It wasn't magic that made the numbers equal, it was an algebraic equation! But that's just the first of many plot lines that we need to tie up before the sun sets on our time together. We've got two more challenge problems to solve, some more algebra to learn, and a math secret agent that appears to be in some sort of trouble. So, with all that in mind, I invite you to turn the page because (after the Final Exam, of course) it's time for the grand finale.

· FINAL EXAM ·

• POLYNOMIALS

1. Categorize each of the following polynomials by the number of terms it contains and its degree, and put it in the proper place in the table below.

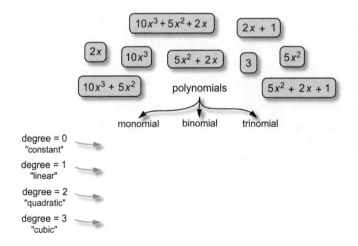

- FUNCTIONS

2. Earlier in the chapter, we looked at the graphs of constant, linear, quadratic, and cubic monomials. But we didn't look at the graphs of monomials with a degree of 4 or higher. So, to see what they look like, plot the degree 4 and 5 monomials

$$y = \tfrac{1}{4} x^4 \text{ and } y = \tfrac{1}{4} x^5$$

You can create these plots on a computer or calculator if you'd like, or by creating a table of values for each function at different values of x and then plotting them here:

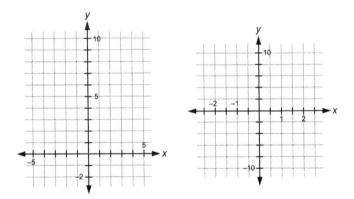

Bonus question: Do you see any similarities between the monomials we've plotted with even degrees and $y = \tfrac{1}{4} x^4$? And, similarly, do you see any similarities between the monomials we've plotted with odd degrees and $y = \tfrac{1}{4} x^5$

- PLOTTING LINES AND OTHER POLYNOMIALS

3. Earlier in the chapter, we looked at the graph of the polynomial $y = x^2 - x - 4$ and noticed that the $-x$ linear term caused the base of the parabola to shift to the right side of the y-axis. Following the same procedure we used earlier, make a plot of the

polynomial $y = x^2 + x - 2$, and compare the result. In particular, which direction is the polynomial shifted relative to the y-axis now?

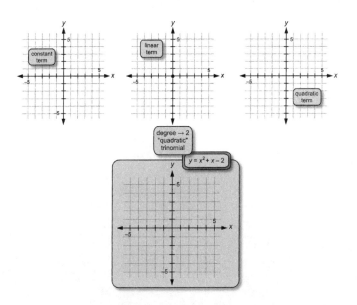

- SYSTEMS OF EQUATIONS

4. Solve the following system of equations by substitution . . . or at least *try* to solve it. Because I guarantee you something "funny" is going to happen. Bonus points if you can figure out what goes wrong!

   $2x + y = 5$ and $4x + 2y = 1$

- SYSTEMS OF INEQUALITIES

5. It's time for an encore version of the monkey and banana story! Since we were last in the jungle, you've learned that the monkeys have free range over all the land defined by the inequality

$y > -\frac{1}{4}(x-1)^2 + 4$. And you've also learned that you have free range over all the land defined by the inequality $y \geq x - 1$. Solve this system of linear inequalities, create a graph showing the solution, and come up with an interpretation about what the solution means in terms of the characters in our monkey and banana story.

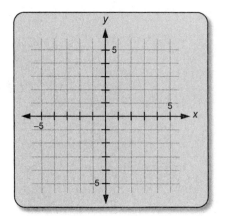

CHAPTER 6

The Root of the Problem

Do not worry about your difficulties in mathematics, I assure you that mine are greater.

—Albert Einstein

If ever math has you down, just think about those words. Because they're pretty much always guaranteed to be true. No matter what troubles you run into, things probably aren't really all that bad. After all, it's probably not like you're stuck trying to unravel the deepest mysteries of the universe, like Einstein was. But in math, as in life, perseverance pays off. And look, perseverance has seen you through to reach the last chapter of this book! Which means both that you've now acquired a great set of algebra skills, and that you're now ready to tackle the more difficult problems we'll look at in this chapter. No, not Einstein difficult . . . but definitely more challenging. And speaking of a challenge . . .

CHALLENGE PROBLEM #2

THE "MYSTERY" PRODUCT

It's time to talk about the second of our challenge problems from the end of the first chapter. Here's how it goes:

If you're told that the product of two consecutive integers is 1056, can you figure out what those two numbers are? And furthermore, do those two numbers have to be positive? Or can one or both of them be negative?

At the time we first saw it, we couldn't solve the problem. And the reason was simple: Because it's an algebra problem . . . and we didn't know algebra yet! But now we do, so let's get started by turning these words into an equation.

TWO UNKNOWNS ↔ TWO EQUATIONS

Okay, the problem says that the number 1056 is the product of two integers. How can we write that? Well, don't think too hard about this . . . it's just as simple as it sounds. The English to equation translation here looks like

$n_1 \bullet n_2 = 1056$

In other words, we've defined two variables, n_1 and n_2, to represent our two unknown numbers, and then we've expressed the relationship we've been told about their product.

So we now have two unknown variables . . . but we only have one equation! In the last chapter, while talking about how to solve systems of equations, we learned that simultaneously solving for the values of two unknown variables requires two equations that each describe different relationships between these variables. Which means that as things stand right now, we're stuck. So, are we out of luck? Well, no . . . this would have made for a lousy problem if we had been. Actually, we've been given more information than we've used so far.

THE SECOND EQUATION

And that important additional piece of information is that the two integers whose product equals 1056 are not just any two random integers . . . they're consecutive integers. And we can express this relationship between the two unknown integers with another equation. Since the word consecutive means "one after another" or "separated by 1," this second equation is just

$$n_2 = n_1 + 1$$

In other words, one of the numbers is larger than the other by one. So we now have two equations that each provide different information about the relationship between the variables n_1 and n_2. Which means that we can now solve the problem. How? Well, as you may have recognized, this is a system of equations—exactly like the systems of equations we studied in the last chapter. And we can apply everything we learned then about solving systems of equations to this problem. In particular, we can use the method of substitution to turn our two equations for two unknowns into a single equation for one unknown.

PUTTING THE PIECES TOGETHER

Okay, since the second equation is already solved for n_2, let's simply take that equation and substitute it in for the variable n_2 in the first equation. When we do that, we turn the equation $n_1 \bullet n_2 = 1056$, into the equation

$$n_1 \bullet (n_1 + 1) = 1056$$

And if we multiply out the left side, we get

$$n_1{}^2 + n_1 = 1056$$

So that's our equation for the variable n_1 . . . but what do we do now? How do we solve it?

✏️ **POP QUIZ: ON A MORE GENERAL NOTE . . .**

Instead of writing an equation that can be used to find two consecutive numbers whose product is 1056, can you write an equation that
can be used to find two numbers whose product is a number we'll call
P and that are separated by an integer amount we'll call S? Write
down the two equations (P and S here aren't unknown variables . . .
they're constants that you get to decide on later), and then combine
them to come up with an equation for one of the numbers.

HOW TO SOLVE IT?

One way to solve this equation is to guess. Seriously, that's a perfectly valid method. For example, you could first guess that $n_1 = 10$.
In that case, the left side of the equation evaluates to $10^2 + 10 = 110$.
But that's a lot less than 1056, which means that the two sides of
the equation are out of balance . . . and that n_1 must be larger.

If you keep doing this for a long time, you will eventually
find the correct value of n_1. In other words, eventually you'll
plug in the right value so that when you evaluate the equation
you'll get $1056 = 1056$—a balanced equation. But this sort of
brute-force method is not a very good idea. For one thing, it's
going to take you a long time. And for another thing, it's not
algebra! And since you now know algebra, there's no reason you
shouldn't take advantage of it.

So, how can we use algebra to solve this problem? Well, this
type of equation is called a "quadratic equation" since it contains a degree 2 polynomial. We've run into equations like it
before (in fact, we plotted a bunch of them in the last chapter),
so they're not brand-new to us. But, unfortunately, we still don't
know how to solve them. Which I guess means that I was a bit
premature in declaring that we were ready to tackle our challenge problems. It's a bummer, I know—although it probably
shouldn't be a big surprise since there's still a big chunk of book
left! So let's move on and see what it has to say.

ALGEBRA DECODER!

- ## QUADRATIC EQUATION

We've talked about the word "quadratic" in relationship to polynomials, and now we've seen it here to describe a type of equation. But what does the word "quadratic" really mean? In fact, doesn't the prefix "quad" mean four? And shouldn't quadratic therefore refer to degree 4 instead of degree 2 polynomials? While this is a perfectly reasonable line of logic, it turns out that the word "quadratic" comes from the Latin word *quadrare* which means something like "to make square." And given that degree 2 polynomials have a term that's squared, the name makes sense after all.

VISUALIZING THE SOLUTION OF THIS QUADRATIC EQUATION

We're stuck . . . at least until we can figure out how to solve problems like $n_1^2 + n_1 = 1056$. In other words, until we can figure out how to solve quadratic equations. In an effort to get unstuck, let's take a look at what a quadratic equation really means . . . and let's do it visually. Here's the picture:

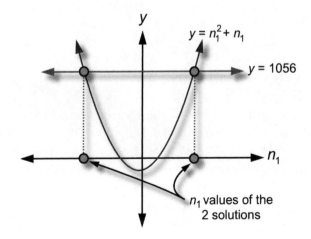

As you can see, there are two different functions plotted on this graph—both of which are polynomials. The first is described by the equation $y = n_1^2 + n_1$, which you'll notice looks very similar to many of the curves we saw in the last chapter. In fact, this function is just a degree 2 binomial and, like all degree 2 polynomials, that means that it's a parabola complete with its characteristic two arms up-in-the-air shape (as long as its numerical coefficient is positive, that is). The second function shown on the graph is the constant $y = 1056$ which, not surprisingly, is a horizontal line.

So what does this all mean? Well, we've looked at solving equations graphically before—way back in chapter 2, so this shouldn't come as a big surprise. But it's a very useful picture to remind yourself of because it tells us what the solution to the equation

$$n_1^2 + n_1 = 1056$$

looks like. No, it doesn't tell us precisely what the solution is (in other words what the actual values of n_1 are), but it does tell us that the equation must have two solutions. And that these n_1 values correspond to the locations where the two functions intersect—one is a positive number, and the other is negative. As we'll see momentarily, those locations are closely related to something called the "roots" of this polynomial.

ROOTS OF POLYNOMIALS

Now we're really getting to the root of the problem, so to speak. Roots . . . like tree roots? No. Okay, like square roots? And cube roots? Well, that's a lot closer. In fact, those types of roots are closely connected to the roots we're talking about . . . but let's hold off on that conversation for now. The roots we're talking about here belong to polynomials. And they play a critical role in helping us find solutions to equations containing them.

WHAT ARE THE ROOTS OF A POLYNOMIAL?

A few paragraphs ago we got stuck trying to solve the polynomial equation

$$n_1{}^2 + n_1 = 1056$$

So now, as the first step in our effort to once and for all get ourselves unstuck, let's use this equation as the stepping-off point for our conversation about roots. In particular, let's start by subtracting 1056 from both sides to get

$$n_1{}^2 + n_1 - 1056 = 0$$

Then let's plot this and see what it looks like

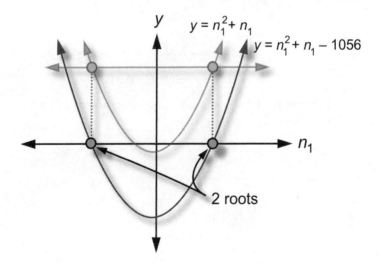

2 roots

As you can see, the function $n_1^2 + n_1 - 1056$ looks just like the function $n_1^2 + n_1$ that we saw before (and which you can also see here), except that it's moved down the y-axis by 1056.

So what's the point of this? Well, as we talked about before, we can find the solutions of the equation $n_1^2 + n_1 = 1056$ by locating the points where the functions $y = n_1^2 + n_1$ and $y = 1056$ intersect. But now by subtracting 1056 we can move the entire function down the y-axis until those two points of intersection fall on the horizontal x-axis. And these two points where the function $n_1^2 + n_1 - 1056$ intersects the x-axis are called the roots of the function.

But why are these roots so important? Well, let's go back to the problem we're trying to solve. We want to know the values of n_1 that make the following equation true:

$$n_1^2 + n_1 - 1056 = 0$$

And, as you can see by looking at the plot we just made, the values of n_1 that solve this equation are precisely the roots of the equation. So the roots *are* the solutions! Now we just need to figure out how to find them.

✏️ **POP QUIZ: WHERE ARE THE ROOTS?**

Locate all roots of the four functions shown below and circle them on the graph.

Cont'd

Cont'd from page 346

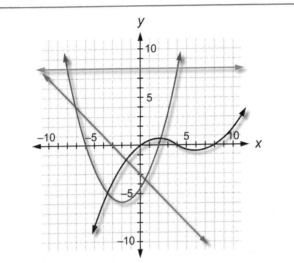

How many roots does each polynomial have and where are they located?

- The constant polynomial has _____ roots.
- The linear polynomial has _____ root located at $x =$ _____.
- The quadratic polynomial has _____ roots located at $x =$ _____ and _____.
- The cubic polynomial has _____ roots located at $x =$ _____, _____, and _____.

Are the roots you found solutions to the following polynomial equations describing the functions? Plug the value of each root into the equation for that polynomial and check.

- Does the root of the linear polynomial solve the equation $-x - 3 = 0$?

        Root 1: _____

- Do the roots of the quadratic polynomial solve the equation $x^2 + 4x - 12 = 0$?

        Root 1: _____

        Root 2: _____

Cont'd

Cont'd from page 347

- Do the roots of the cubic polynomial solve the equation $x^3 - 12x^2 + 32x = 0$?

  Root 1: _____

  Root 2: _____

  Root 3: _____

Assuming the roots solved the equations (which they had better), what does it mean? Well, it means that we can find the solution to an equation like "some polynomial" $= 0$ just by plotting the polynomial and finding the location or locations where it crosses the x-axis. Or, perhaps we can find an even more clever way to find roots . . .

Rest assured that we will eventually come back to solving the equation from our challenge problem (it may be when you least expect it), but now we can see that there are a number of things we need to learn before we can do it. And first up on that list is figuring out how best to find the roots of a polynomial. So let's put the challenge problem on hold for a while, and start working through those prerequisites.

THE FUNDAMENTAL THEOREM OF ALGEBRA

In the last Pop Quiz, we saw that different polynomials have different numbers of roots. Which leads one to wonder if there is some method to the madness. It can't all just be random, right? To see, let's investigate the relationship between the degree of a polynomial and its roots.

As you'll recall, the degree of a single variable polynomial (which is the type we'll be dealing with here) is just equal to the largest power of the variable in any of its terms. For example, a constant polynomial has a degree of 0, a linear polynomial has a degree of 1, a quadratic has a degree of 2, and so on.

And it turns out that the degree of a polynomial is *very* closely related to the number of roots it has. In fact, this idea is the basis for something called the "fundamental theorem of algebra." And with a name like that, you know it must be important:

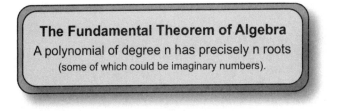

The Fundamental Theorem of Algebra
A polynomial of degree n has precisely n roots
(some of which could be imaginary numbers).

We'll come back to talk about the statement in parentheses about imaginary numbers in a few minutes . . . so go ahead and ignore that for now.

POP QUIZ: HOW MANY ROOTS?

Determine the number of roots that each of the following polynomials must have:

1. $x^2 \bullet (x+1)$ . . . . _____
2. $x^3 - 2x^2 + x + 3$ . . . . _____
3. $(x^2 + x) \bullet x^2$ . . . . _____
4. $(x+1) \bullet (x+1)$ . . . . _____

Hint: You'll need to distribute and use FOIL to fully expand some of these so that you can see all of the terms of the polynomial before determining how many roots it has. And don't forget that the number of roots is equal to the degree of the polynomial.

REDUCIBLE POLYNOMIALS

Let's check and see if the statement that a polynomial of degree n has precisely n roots makes sense. If you count the number of times each of the polynomials plotted on the following graph crosses the x-axis, you'll see that a constant polynomial (degree = 0) has no roots, a linear polynomial (degree = 1) has one root, a quadratic polynomial (degree = 2) has two roots, and a cubic polynomial (degree = 3) has three roots.

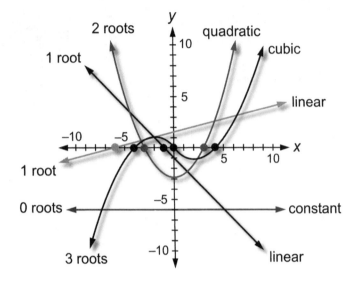

So the rule about number of roots and degrees seems to hold up. All of the polynomials shown on this graph are what's called "reducible over the real numbers." In other words, any polynomial that intersects or crosses the x-axis at least one time is a reducible polynomial. But you might be wondering why we're bothering to give this type of polynomial a specific name. Do we need a name to distinguish them from some other type of polynomial? Yes.

IRREDUCIBLE POLYNOMIALS

The reason for calling polynomials like the ones we just looked at "reducible" is best explained by looking at a plot of some polynomials that are not reducible (at least not as far as we're concerned). And lucky for us, that's exactly what we have here:

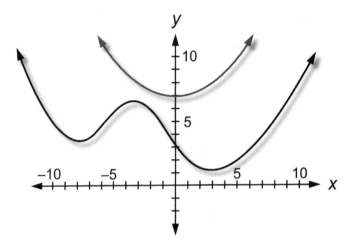

So what's the big deal? Well, if I ask you what the roots of these polynomials are, would you be able to tell me? No, because these functions don't intersect or cross the x-axis . . . and that's how we've been finding roots. Functions like these that don't cross the x-axis are called "irreducible."

Why does this matter? Well, it has to do with our rule that the degree of a polynomial is equal to how many roots it has. As you can see, that rule seems to be broken here. The top curve looks like a quadratic, and the bottom squiggly polynomial must have an even higher degree than the quadratic. Yet neither of them cross the x-axis . . . so it seems that neither of them have any roots. And that puts our rule in serious jeopardy. But despite appearances, these polynomials really do have roots— and our rule really does hold for them too. It's just that these roots aren't made of real numbers like the kind you're thinking of. These roots are imaginary.

OVERACHIEVER  BADGE

IMAGINARY ROOTS

The roots of some polynomials are actually imaginary numbers. We're not going to go into too much detail about imaginary numbers here, but to get an idea of where these roots come from, let's look at the polynomial $x^2 + 1$:

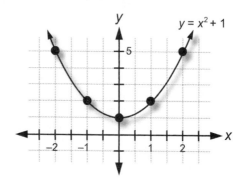

As you can see, at no point is the polynomial $x^2 + 1$ ever equal to zero. In other words, the polynomial never crosses the x-axis. But does that really mean there's *no* solution to the equation $x^2 + 1 = 0$? Actually, there is . . . you just have to wander into the world of imaginary numbers to find it. The trick is to create a new number, i, that satisfies the equation

$$i^2 = -1$$

Then, if we set $x = i$ in the equation $x^2 + 1 = 0$, we get

$$i^2 + 1 = -1 + 1 = 0$$

Cont'd

Cont'd from page 352

Which means that we *can* actually solve the equation $x^2 + 1 = 0$, we just don't really understand how to do it yet. And the good news is that when we include these imaginary roots, the rule that the degree of the polynomial is equal to the number of roots is correct for irreducible polynomials too. One word of warning: some polynomials have *both* real and imaginary roots. So this stuff can get pretty complicated. But you don't need to worry about that for now.

POP QUIZ: **REDUCIBLE OR IRREDUCIBLE?**

Which of the polynomials shown in the graph below are "reducible over the real numbers"? Which are irreducible?

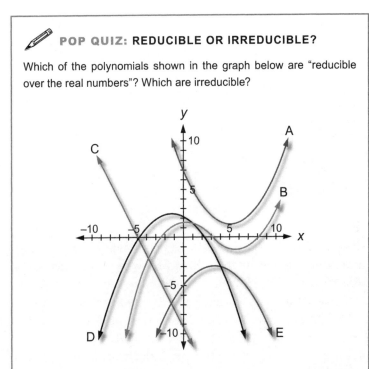

Remember, a reducible polynomial intersects the *x*-axis and therefore has roots that are real numbers. An irreducible polynomial does *not* cross the *x*-axis and therefore only has roots that are imaginary numbers.

MULTIPLICITY

So despite that brief scare, our rule that says the degree of a polynomial is equal to the number of roots it has is still holding up . . . or is it? What about polynomials like x^2 and x^3? If you plot those polynomials, you'll see that they each intersect the x-axis only one time:

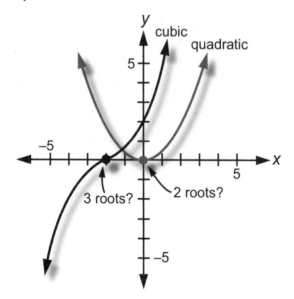

But x^2 and x^3 are degree 2 and 3 polynomials, which according to our rule means they should have two and three roots. What's going on here? Well, you might find this hard to believe since everything you're seeing is telling you otherwise, but x^2 and x^3 really do have two and three roots in this plot—you just can't see them because they're all identical! In other words, the multiple roots of each function are located at exactly the same point on the x-axis, and that makes it really hard to see them all.

To see what's really going on here, let's write the polynomial x^2 as $(x-0) \cdot (x-0)$. I know this seems like a strange thing to do, but something interesting happens when we do it. Notice that we can use FOIL to multiply out $(x-0) \cdot (x-0)$ and end up right back at x^2, which means that everything we've done so far is okay. Now, let's use the fact that $x^2 = (x-0) \bullet (x-0)$ to write the polynomial equation $x^2 = 0$ as

$$(x-0) \bullet (x-0) = 0$$

How can we solve this equation? Well, we can set $x = 0$, and we'll end up with $0 = 0$. But notice that we could set $x = 0$ in *either* of the two parts of the expression on the left and get $0 = 0$. So there are actually two different ways to solve this problem. Which means that despite all appearances the polynomial x^2 really does have two roots—they just both happen to be zero. There's a word to describe roots like this—we say that the root $x = 0$ of the polynomial x^2 has a "multiplicity" of two.

 **POP QUIZ:**
WHAT'S THE MULTIPLICITY OF EACH ROOT?

Determine the locations of the roots of the following polynomials and their multiplicity. The first problem is done for you as an example. If you need help figuring out how this works, take a look at the previous paragraph where we went through the same type of example for the polynomial x^2. And you can always look at the Math Dude's Solutions at the back of the book for an explanation.

1. $x^2 - 4x + 4 = (x-2) \bullet (x-2)$ ← You can use FOIL to see that this is true.
    Root 1: Located at $x = \underline{\ 2\ }$ with multiplicity $\underline{\ 2\ }$.

2. $x^3 - 2x^2 + x = (x-0) \bullet (x-1) \bullet (x-1)$
    Root 1: Located at $x = \underline{\ \ \ }$ with multiplicity $\underline{\ \ \ }$.
    Root 2: Located at $x = \underline{\ \ \ }$ with multiplicity $\underline{\ \ \ }$.

3. $x^3 - x^2 - x + 1 = (x-1) \bullet (x+1) \bullet (x-1)$
    Root 1: Located at $x = \underline{\ \ \ }$ with multiplicity $\underline{\ \ \ }$.
    Root 2: Located at $x = \underline{\ \ \ }$ with multiplicity $\underline{\ \ \ }$.

REDUCIBLE VS. IRREDUCIBLE POLYNOMIALS

We've covered a lot of ground, so let's take a minute to recap exactly what we've learned so far about polynomials and their roots. First, the roots of a polynomial are nothing more than the solutions to the equation you get when you set that polynomial equal to zero. That's always true, independent of whether or not the polynomial is **reducible** or **irreducible**.

Reducible polynomials have roots that are real numbers. For example, the polynomial $x^2 - 9$ has two roots, $x = 3$ and $x = -3$, as you can verify by solving the equation $x^2 - 9 = 0$. For reducible polynomials, you can find the roots graphically by looking at the locations where the polynomial intersects the x-axis.

Irreducible polynomials have roots that are imaginary numbers (which you don't really need to worry about understanding at this point). For example, the polynomial $x^2 + 9$ does not have roots that are real numbers, but it does have two imaginary roots. If you look at the graph of an irreducible polynomial, you'll see that it never intersects the x-axis.

We also learned that the number of roots that a polynomial has is equal to the degree of the polynomial. Some polynomials appear to buck this trend because they have roots that occur more than once. The number of times a root occurs for a polynomial is called the **multiplicity** of that root. And once we include the multiplicity of all roots, and we include the imaginary roots for irreducible polynomials, the statement that the number of roots is equal to the degree of the polynomial is *always* true.

WRITING POLYNOMIALS IN FACTORED FORM

It's now time to put together everything we've learned so far about polynomials so that we can take the next big step toward finally being able to solve the equation from our challenge problem that we got stuck on. Earlier in the chapter we talked about the "fundamental theorem of algebra" which says that the degree of a polynomial is equal to the number of roots it has. And now that we understand what this means for the different types of polynomials, we're ready to put it to good use.

One consequence of the theorem is that *any* polynomial can be written in what's called a "factored" form that looks like this:

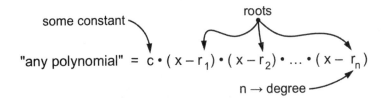

In words this says that any polynomial can be written as a product of some constant and a number of what are called "factors of the polynomial." Each factor looks like $(x - r_i)$, and all of these things called r_i are just the roots we've been finding. The subscript "i" here is used to label which root we're talking about (it has nothing to do with the imaginary i that we talked about before). So r_1 is the first root, r_2 is the second, and so on. As you can see, the equation for "any polynomial" is a product of n of these factors, where n is the degree of the polynomial.

Another way to say all this is that we can build a polynomial by multiplying its factors—which are made from its roots—together. And once you understand that, then you can finally see how the roots of a polynomial are related to roots of numbers that we talked about before . . . things like square roots, cube roots, and so on. The square root of a number tells you what number you need to multiply together to make that number. And similarly, the roots of a polynomial tell you what you need to multiply together to make that polynomial. So the two terms really are connected after all!

At first glance, it seems pretty amazing that any polynomial can be written as the product of its factors. Especially when you realize that each of those factors is simply an equation for a line.

Which says that any polynomial—no matter how complicated—can be built just by multiplying together a bunch of lines! Pretty awesome, right?

It's hard to imagine that it really works. But fortunately for us, there is a way that we can see that it does: We can make a graph showing a couple of lines, multiply those lines together, and then see the resulting polynomial. In fact, let's do that for the polynomial $x^2 - 1$, which I happen to know can be written in factored form as $(x + 1) \cdot (x - 1)$:

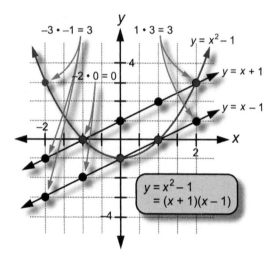

How does this work? Well, all you have to do is pick some points along the x-axis, and then multiply the values of the lines $x + 1$ and $x - 1$ at those points. The numbers you get are the values of the polynomial $x^2 - 1$ at those points. Do this for enough points, and you can create a parabola from nothing more than a pair of lines!

POP QUIZ:
PICTURING POLYNOMIALS IN FACTORED FORM

Let's create a graph of the polynomial created by multiplying the three linear polynomials x, $x+1$, and $x-1$ together. First, plot these three lines. Then pick some points along the x-axis and calculate the product of the values of all three lines at those locations. Finally, plot these points to create your new polynomial. What degree will this polynomial be?

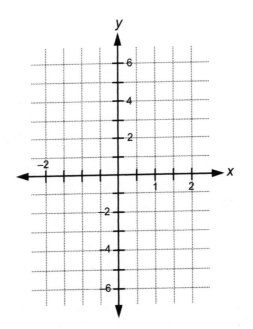

Bonus question: Can you figure out which polynomial you are creating? The distributive property and FOIL will help you here.

FACTORING POLYNOMIALS

So here's the deal: Now that we've seen that we can solve equations with polynomials simply by finding the locations of their roots, and that it's possible to write any polynomial as a sequence of factors that give us the locations of these roots, it makes sense for us to figure out how to turn any polynomial into just such a sequence of factors. Why? Because once we've done that, we've solved the problem. So, it's time for us to learn how to factor.

FACTORING NUMBERS

But before we start factoring polynomials, let's take a look at a simpler but related type of factoring that should be more familiar to you—factoring numbers. The goal when factoring a number is to find its factors. Yes, that's quite logical, but not terribly informative until you remember that the factors of something (be it a number or a polynomial) are just the things that you can multiply together to make that something.

So the goal of factoring a number is to figure out which whole numbers you can multiply together to get the number you're factoring. For example, here are all the ways to factor the number 18:

$$18 \cdot 1 = 18$$
$$9 \cdot 2 = 18$$
$$6 \cdot 3 = 18$$
$$3 \cdot 3 \cdot 2 = 18$$

Each of these products give 18, which means that each of the products on the left are a valid factorization of the number 18. And, therefore, the numbers on the left are all factors of 18. But with numbers, we're usually not interested in knowing all the factors. Often we're only interested in knowing factors that are prime numbers—called prime factors.

ALGEBRA DECODER!

• PRIME NUMBER

Prime numbers are very special positive integers that are only evenly divisible (meaning without fractional pieces left over) by the two numbers 1 and the number itself. The first prime number is therefore 2, the next is 3, then 5, 7, 11, 13, and on and on . . . there are an infinite number of them.

So of all the factorizations of 18 above, only the last one is the prime factorization of 18. Why? Because that factorization contains nothing but prime numbers: 3, 3, and 2. And, because there can only be one prime factorization for a given number.

You probably learned to factor numbers using a technique that looked something like this:

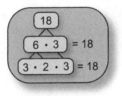

The idea is to start with the number you're factoring at the top, and then continue to break this number apart into products of successively smaller numbers. If you keep doing this, you'll eventually find that you can't break any of the numbers apart anymore. And that means that you're done since you now have a list of nothing but prime numbers. In other words, this is the prime factorization.

This is one of those things in math that seems really simple (especially because most of us learned to do it when we were young), but is actually quite remarkable since it shows that every single one of the infinite number of positive integers can be built by multiplying prime numbers. This means that you can think of prime numbers as fundamental building blocks from which all other positive integers can be created!

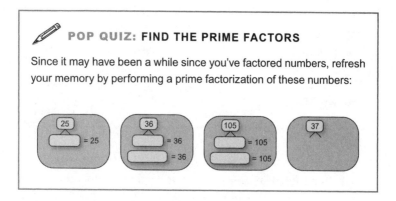

POP QUIZ: **FIND THE PRIME FACTORS**

Since it may have been a while since you've factored numbers, refresh your memory by performing a prime factorization of these numbers:

GREATEST COMMON FACTOR OF A GROUP OF NUMBERS

The greatest common factor (often abbreviated GCF) is another one of those things you typically learn to find in school . . . and then proceed to forget about. But it's going to be very helpful to us when we start factoring polynomials in a minute, so let's refresh our memories

The idea behind the greatest common factor is simple—in fact, the name does a pretty good job at describing what it is. The greatest common factor between two or more numbers is just the largest number that evenly divides into those numbers.

Here's a way for you to think about the greatest common factor that should help you keep track of what's going on. Each of the following sets of circles (which I'll call "factorization circles") represents the factorization of a number. You can find the number being factored at the center of each circle. The ring around that number contains all the possible first steps in the factorization of that number.

For example you can see that there are three different ways to break apart the number 12 on the top right: 1 • 12, 3 • 4, and 6 • 2. You can then continue factoring along one of these paths by adding another ring on the outside into which you break apart the numbers further. And if you keep adding rings and breaking apart the numbers until you can't break them apart anymore, you'll have found the prime factors of the number in the middle.

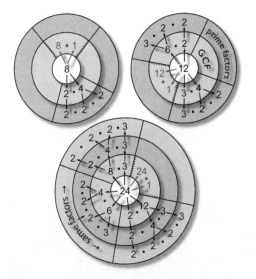

But finding prime factors isn't really why I'm showing this to you. The main advantage to this way of looking at factoring is that it makes it very easy to find the greatest common factor between two or more numbers. All you have to do is create a factorization circle for each number, and then find the largest number in common from the first ring of each circle.

For example, to find the greatest common factor of the three numbers in the factorization circles drawn above—8, 12, and 24—just look for the largest number in common from the first ring of each of the three circles. Since 8 has the fewest numbers in its first ring, it makes sense to start there to look to see which of those numbers appear in the other first rings. The biggest number in the first ring of 8 is (not surprisingly) 8. While that does occur in the first ring of 24, it's not in the first ring of the factorization circle for 12. So, that means we need to move on to the next biggest number in the first ring of 8, which is 4. And since 4 appears in the first ring of all three numbers, it's the greatest common factor!

Notice that you don't actually have to fill out the entire factorization circle to use it to find the greatest common factor. All you actually have to do is fill out the first ring of each circle and you're good to go.

POP QUIZ:

FIND THE GREATEST COMMON FACTOR OF A GROUP OF NUMBERS

As with prime factors, it may have been a while since you've found a greatest common factor. So, to refresh your memory and give you a chance to play with a few factorization circles, find the GCF of these numbers:

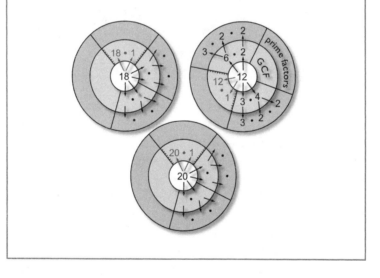

GREATEST COMMON FACTOR OF A GROUP OF MONOMIALS

We're now ready to leave the world of numbers and dive into the world of factoring polynomials. And the first step in doing that is to figure out what the greatest common factor looks like in this world. Why is it any different than for numbers? Well, in one way it isn't any different—the basic goal is the same. But in another way it is different since we're dealing with variables!

For example, when dealing exclusively with numbers, a problem of finding the greatest common factor might sound like: "Find the value of the GCF for the numbers 32, 54, and 18." But in the world of polynomials, a similar problem might sound like: "Find the GCF of the monomials $10x^3$, $5x$, and $20x^2$." Although these sound *very* different, they're really not. In fact, the exact same factorization circle that we used with numbers is helpful here too.

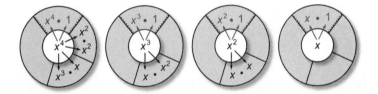

These are all the possible ways of factoring the monomials created by raising the variable x to powers from 1 to 4. I've only shown the first rings of the factorization circles here since we're only interested in looking at the greatest common factors. But you could continue to add rings and factor further if your goal was to find the "prime factorization" (which for x^4 is just $x^1 \bullet x^1 \bullet x^1 \bullet x^1$).

As you can see on the left, there are three possible ways to factor the monomial x^4. The first way shown on the top, $x^4 \bullet 1$, is kind of trivial, but it could end up giving the greatest common factor between multiple monomials. The other two ways of factoring x^4 are $x^2 \bullet x^2$ and $x^3 \bullet x$. Which means that the monomial x^4 has factors of x, x^2, x^3, and x^4. So what's the point of this?

Well, just like finding the greatest common factor of numbers, we can use these monomial factorization circles to figure out the greatest common factor among just the variables in a list of monomials. For example, if we're asked to find the GCF of the monomials $10x^3$, $5x$, and $20x^2$, we could proceed by splitting the problem into two parts:

1. Find the GCF of the numbers 10, 5, and 20 as outlined in the last section. Answer=5.
2. Find the GCF of the variables x^3, x, and x^2 by finding the greatest common power in the first rings of the factorization circles of these three monomials. Answer=x.

Q&DT ▶ The complete greatest common factor of these three monomials is therefore the product of these two things. And the answer is $5x$. So the greatest common factor of a group of monomials is just the product of the GCF of the numerical constants in the monomials and the greatest common power among the variables in the monomials.

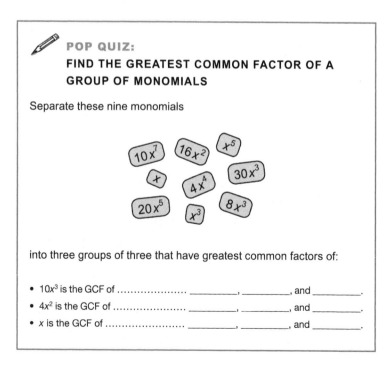

POP QUIZ:

FIND THE GREATEST COMMON FACTOR OF A GROUP OF MONOMIALS

Separate these nine monomials

$10x^7$ $16x^2$ x^5

x $4x^4$ $30x^3$

$20x^5$ x^3 $8x^3$

into three groups of three that have greatest common factors of:

- $10x^3$ is the GCF of .... _____, _____, and _____.
- $4x^2$ is the GCF of .... _____, _____, and _____.
- x is the GCF of .... _____, _____, and _____.

FROM GREATEST COMMON FACTOR
TO FACTORING

Now that we know how to find a greatest common factor—both of a group of numbers and of a group of monomials—we're ready to start factoring polynomials. As we've discussed, the goal of factoring a number is to represent that number as a product of factors. And the same thing is true for polynomials

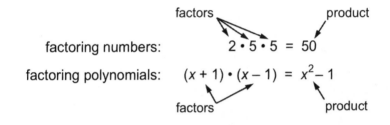

factoring numbers: $2 \cdot 5 \cdot 5 = 50$

factoring polynomials: $(x + 1) \cdot (x - 1) = x^2 - 1$

Little did we know it at the time, but back in chapter 4 when we were learning about the distributive property and how to FOIL, we were actually factoring backwards. So the whole goal of the next section is ultimately to figure out how to undo what we did back then. Does that mean we did something wrong or undesirable before? Not at all. Some problems call for one thing, and others call for the other. It's yin and yang. A most harmonious balance.

We can sum up this complementary relationship between distribution and factoring by saying that distribution is the process of turning a product into a sum, and factoring is the process of turning a sum into a product. For example, here's how these two processes work back and forth in the distributive property:

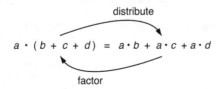

$$a \cdot (b + c + d) = a \cdot b + a \cdot c + a \cdot d$$

and here's how they work when we extend the distributive property to act on a pair of binomials—also known as FOIL:

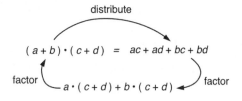

So, as you can see, all that we have to do now is start un-FOILing. But we won't just start trying to do it willy-nilly; there are a few guidelines and tricks that will make your factoring experience much more pleasant.

M A T H   B R A I N   G A M E

IDENTIFY THE POLYNOMIAL IDENTITY

The first item for you to tuck into the bag of tricks you're putting together to help simplify your polynomial factoring life are a few very handy identities. What's an identity? The word "identity" can have several meanings in math, but in this case it's referring to something that tells you about a relationship between two expressions. Isn't that just an equation? Well, yes . . . they're very similar. The only difference is that an identity is true no matter what values you put in for the variables—which certainly isn't true for an equation like $x + 1 = 2$. But let's not dwell on the finer points of mathematical language right now, let's get on with learning about these identities.

UNIDENTIFIED IDENTITIES

Of course, since this is a "brain game," I'm not just going to give away these identities . . . at least not yet. I first want you to stretch your brain a bit by working toward figuring out a few things that will give you a better feel for what the identities we're going to learn mean. In particular, can you figure out which of the two identities below go with each of the accompanying drawings depicting the meaning of the identities?

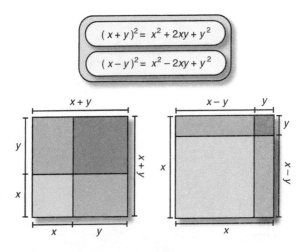

Here's a hint: One of these drawings looks *very* similar to something we looked at in chapter 4 (think FOIL). And if you can figure out which that is, then you'll be able to figure out which the other is too. Think about it, and then read on to find out which identity goes with which drawing, and to get an explanation of how these drawings depict the geometrical meaning of the identities.

IDENTITY ONE: THE SQUARE OF THE SUM OF TWO VARIABLES

So, did you figure out which identity looks very similar to something we've seen before? Well, back when we talked about FOILing, we wrote down an equation like this:

$$(a+b) \bullet (c+d) = ac + ad + bc + bd$$

But if we set $a=x$, $c=x$, $b=y$, and $d=y$, then this just becomes

$$(x+y)^2 = x^2 + xy + xy + y^2 = x^2 + 2xy + y^2$$

And that's exactly the same as the first identity above. The drawing below shows how to interpret the geometric meaning of each term in this multiplied out identity. The basic explanation is that the area of the large square is equal to the sum of the areas of the four smaller squares, which is exactly what we saw when we derived FOIL in chapter 4.

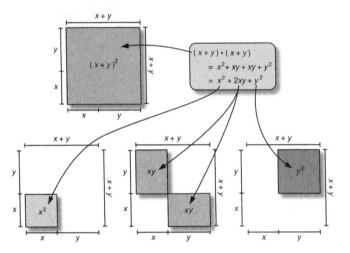

IDENTITY TWO: THE SQUARE OF THE DIFFERENCE OF TWO VARIABLES

So now you know that the second identity for $(x-y)^2$ must go with the drawing on the right from earlier. The drawing that follows describes how to interpret each piece of this slightly more complicated multiplied out identity. As you can see, the identity essentially says that a smaller square can be created from a bigger square by subtracting off the edges. Pay careful attention to whether terms are being added or subtracted to ensure that you see how the geometric shapes are either being added or taken away to make the smaller square.

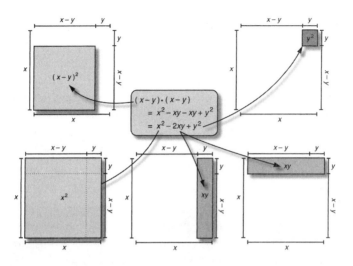

A FEW MORE IDENTITIES

To round out your collection of identities, here are three others that are sure to prove worthy of inclusion in your bag of tricks. We'll take a look at the geometrical interpretation of the first of these identities in the Final Exam.

◀ Q&DT

Identities Useful for Factoring Polynomials

Difference of squares: $x^2 - y^2 = (x + y) \cdot (x - y)$

Square of the sum: $x^2 + 2xy + y^2 = (x + y)^2$

Square of the difference: $x^2 - 2xy + y^2 = (x - y)^2$

Sum of cubes: $x^3 + y^3 = (x + y)(x^2 - xy + y^2)$

Difference of cubes: $x^3 + y^3 = (x - y)(x^2 + xy + y^2)$

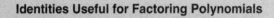

HOW TO FACTOR POLYNOMIALS

It's now time for the main event . . . the moment you've been waiting for. It's time to put everything that we've learned so far together with all of our tips, tricks, and tools to start doing some honest to goodness factoring of polynomials.

Let me start with a word of wisdom: Factoring can be hard . . . especially when you're just getting started. So if you feel like it's hard, just know that you're not alone. But we're going to do everything we possibly can to make it as easy as possible—that's why we've spent all this time getting ourselves ready.

And although factoring may be hard now, rest assured that after you've been doing it for a while, it will get easier. Factoring is definitely an area of algebra where practice pays off. So we'll do our fair share of that. But having a good playbook to guide you through the process is just as important. So, with all that in mind, let's get this tutorial started with an overview of the four-step plan for factoring polynomials that we're going to follow:

Q&DT ▶

> **The 4 Steps to Successful Factoring of Polynomials**
> 1. Factor out the greatest common factor.
> 2. Check if anything can be factored using one of our identities. If so, do it!
> 3. Factor any quadratic trinomial parts.
> 4. Return to the beginning and start again. Factor until there's nothing left to factor!

While going through this tutorial, we're going to work on factoring five different polynomials (meaning you're going to become very friendly with these guys). Here are your new best friends:

1	$2x^3 + 2x^2 - 12x$
2	$6x^2 - 2x$
3	$x^3 + 4x^2 + 4x$
4	$12x^6 - 27x^2$
5	$8x^2 + 20x - 12$

■ STEP 1: FACTOR OUT THE GREATEST COMMON FACTOR

The first step in factoring polynomials is to look for and factor out the greatest common factor. As we talked about earlier, the easiest way to do this is to split the problem up into two pieces. First, find the greatest common numerical factor of all the numerical coefficients. Then find the greatest common power among the variables in each term of the polynomial. The greatest common factor for the polynomial is then the product of these two parts. You can use the factorization circles method that we looked at earlier to help you, although after you practice enough you'll be able to do the work in your head.

Now let's apply our method for finding the greatest common factor to the polynomials we're working on. The numerical coefficients in the first polynomial are 2, 2, and 12. Since $12 = 6 \cdot 2$, I can immediately see that the greatest common numerical factor for this polynomial is 2. How about the greatest common power of the variables? Well, all three terms have at least one power of x, so the greatest common power of the variables for this polynomial is just x. Multiplying these together, we see that the greatest common factor for the first problem is $2x$.

But we're not done; we now need to factor $2x$ out of the polynomial. And the easiest way to do that is to think of the problem like this:

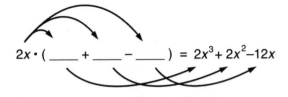

$$2x \cdot (\underline{} + \underline{} - \underline{}) = 2x^3 + 2x^2 - 12x$$

In other words, you need to treat this like you're using the distributive property backwards. For the first term, you know that you are multiplying $2x$ by some unknown, and that this must give you $2x^3$. If you can figure out the unknown in your head, then that's great. But if you need to use a little algebra here to help you out, you can do that too. How? Well, if we solve this little problem

$$2x \cdot y = 2x^3$$

for the unknown y, then we'll know what to stick in for the first unknown term. If you work this out, you'll find that the answer is just $y = x^2$. So at this point, our problem looks like

$$2x \cdot (\underline{x^2} + \underline{} - \underline{}) = 2x^3 + 2x^2 - 12x$$

Now you need to go through the same sequence of steps to find the unknown values for the last two terms. Go ahead and try that now. You should end up getting this:

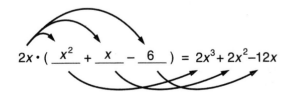

$$2x \cdot (\underline{x^2} + \underline{x} - \underline{6}) = 2x^3 + 2x^2 - 12x$$

Which means that you've factored out the greatest common factor and found that $2x^3 + 2x^2 - 12x = 2x \bullet (x^2 + x - 6)$. We're done with this polynomial for the time being. But before moving on to the next step, take a shot at factoring out the greatest common factor from each of the other four polynomials, and finish filling this table out with your results:

If everything has gone well, you should now be completely finished factoring one of the five polynomials. Which one? The second one. How do you know that you're done? Well, when all the terms look like "$(x - \text{constant})$" as we talked about before (where the constant here is related to the root of the polynomial), you know you're done. But a safer way to really make sure you're done is to check through all of the steps to see if you can use any of the other techniques to factor more. If you can't, then you really know that you're done. So that's it for number two. For the rest of the polynomials, let's keep on truckin'.

■ STEP 2: FACTOR USING IDENTITIES

The second step in our four-step process is to check and see if any of the identities we learned in the last Math Brain Game can be used to further factor the polynomial. When they can, it's very good news—because most of the time that means you're finished with the problem. For this step of the tutorial, I'm going to walk through solving the fourth polynomial, since I know that we can factor that one using an identity. So, here's where we stand right now:

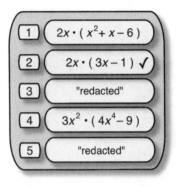

This table now shows the current state of the first problem after we worked through the first step together, the second problem which is now fully factored, and the answer that you should have gotten for the fourth problem after the end of the first step. The other two problems have been "redacted" until we're ready to deal with them (in other words, I want you to think about and work on them by yourself for now).

Okay, so how do you go about figuring out if you can use one of the identities or not? Well, sorry to say it, but there's no magic answer. You just have to look through them all and think a bit to see if they fit the bill. This is one of the parts of factoring that gets much easier as you practice more. Eventually, you'll memorize the identities, and then you'll begin to recognize when a polynomial can be factored using one of them.

In this case, I've noticed something about the fourth problem that tipped me off to the fact that one of the identities can help us here. But before I get to that, let me mention that although we're not done with the fourth problem, we are done with the first part of it—the $3x^2$ part—since that part is fully factored. Which means that we only need to work on the part of the problem in parentheses.

So, what identity can we use here? Well, it's not immediately obvious, but if we rewrite $4x^4 - 9$ like this

$$(2x)^2 - 3^2$$

then the answer becomes much clearer. Take another look at the identities. In particular, take a look at the "difference of squares" identity:

$$x^2 - y^2 = (x+y) \bullet (x-y)$$

As you can see, the expression $(2x)^2 - 3^2$ is indeed a difference of two squares, which means that it can be factored like this

$$(2x+3) \bullet (2x-3)$$

Again, there's no magic to figuring out that this problem could be factored as a difference of squares. It's really just a combination of being alert to the possibility and experience.

So, after that bit of factoring, the fourth problem is now finished. Which means we've got two down . . . and three to go. Before moving on the the next step, try and see if any of the other problems can be factored using one of the identities. I'm betting one of them can be, but I'm going to leave it up to you to figure out which one . . . and to do the factoring. Once you've figured that out, then we'll actually have finished three of the five problems. Of course, you can always check the Math Dude's Solutions at the end of the book to make sure you're on the right track.

Here's the current state of affairs:

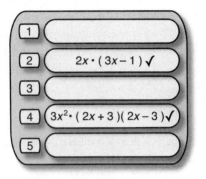

Try to factor the other problems using the identities. And when you're finished with that, move on to the next step where we'll finish factoring the two polynomials that are still going to need some work.

■ STEP 3: FACTOR QUADRATIC TRINOMIAL PARTS

This step is where a lot of the action happens. Why? Well, to be honest, it's not very often that one of the identities actually works. When they do, it's great—but the majority of problems that you'll have to factor just aren't that special. Which means that after factoring out the greatest common factor in the first step, you'll often end up at this step staring at a quadratic trinomial that looks something like $ax^2 + bx + c$ to factor. So how do we factor them?

Well, let's come at that question from the other direction. Let's start by using FOIL to create a quadratic trinomial, and then let's take a look at the result to see what we can learn about how quadratics factor:

$$\text{sum}$$
$$3 + -5 = -2$$

$$(x + 3)(x - 5) = x^2 - 2x - 15$$

$$3 \cdot -5 = -15$$
$$\text{product}$$

As you can see, we've multiplied together $(x+3)$ and $(x-5)$ to get $x^2-2x-15$. But the interesting thing isn't what we got, it's that the *sum* of the two numbers on the left (which are actually the roots) end up equaling the numerical coefficient of the linear term on the right. And the *product* of the two numbers on the left end up equaling the final constant term on the right. Which means that we now have a pattern that we can use to help us figure out how to factor quadratic trinomials.

In order to proceed, let's split the world of quadratics up into two types . . .

TYPE ONE: QUADRATIC COEFFICIENT = 1

First, let's look at quadratics that have a numerical coefficient of 1 on the quadratic term. In other words, these are quadratics that look like x^2+bx+c. To see how this works, let's take a look at the quadratic trinomial portion from the first polynomial on our list. As you'll recall, after factoring out the greatest common factor, we were left with a quadratic part of the polynomial equal to x^2+x-6. So here's the game plan. Given what we learned earlier about the sum and product of the roots of the two factors, we now know that the factorization of this polynomial has to look like

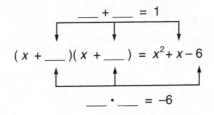

So now we just have to solve the puzzle of figuring out which two numbers add to 1 and multiply to −6. Since there are an infinite combination of numbers that add to 1, that's not a very good place to start. But the list of pairs of numbers whose product is −6 is much more manageable. In fact, there are four options:

- $-1 \bullet 6 = -6 \rightarrow$ number pair of -1 and 6
- $1 \bullet -6 = -6 \rightarrow$ number pair of 1 and -6
- $-2 \bullet 3 = -6 \rightarrow$ number pair of -2 and 3
- $2 \bullet -3 = -6 \rightarrow$ number pair of 2 and -3

Now all we have to do is figure out which of these four pairs of numbers adds up to 1. And pretty quickly we see that it must be the pair of numbers -2 and 3. Which means that the final factorization of $x^2 + x - 6 = (x - 2)(x + 3)$.

Okay, four of our five polynomials are now factored! Just one more to go . . .

TYPE TWO: QUADRATIC COEFFICIENT ≠ 1

Which brings us to the second type of quadratic—those that have a quadratic term with a numerical coefficient of something other than 1. In other words, these are quadratics that look like $ax^2 + bx + c$. Why did we break these off from the others? Because they're a little harder to solve. And it makes sense to do the easy ones first to get the general idea.

To see how to deal with these guys, let's take a look at the quadratic trinomial portion of our last polynomial—yes, it's time to remove it from the "redacted" list. After factoring out the greatest common factor, the remaining quadratic portion of the polynomial is $2x^2 + 5x - 3$. Since the numerical coefficient on the quadratic term is no longer 1, our game plan has to change a bit:

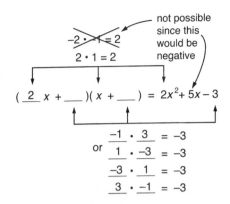

As you can see, things get more complicated when the leading numerical coefficient isn't 1. Let me walk you through a bit of the logic with factoring a problem like this. First, the nice pairing that we had before with the sum and product of the numbers on the left equaling some of the coefficients on the right is gone. However, it is still true that the product of the two unknown numbers on the left must equal the constant on the far right of the quadratic. So the number in the spaces on the left must be either –1 and 3, or 1 and –3. Also, there must be a 2 preceding an x in the factors on the left in order to also get the 2 on the right side of the equation.

Putting this all together, we see that there are four possible factorizations that are consistent with the conditions we just described:

- $(2x-1)(x+3)=2x^2+5x-3$
- $(2x+1)(x-3)=2x^2-5x-3$
- $(2x-3)(x+1)=2x^2-x-3$
- $(2x+3)(x-1)=2x^2+x-3$

At this point, we can simply multiply out all the possibilities and determine which is correct. In this case, it was the first one. So the final factorization of $2x^2+5x-3=(2x-1)(x+3)$.

Eventually, after you practice factoring more, you probably won't have to write out all the possibilities like this—you'll be able to do a lot of the work in your head. But for now it's good to write everything out since that will help you begin to learn how the various pieces end up fitting together.

Okay, we've now fully factored all of the polynomials on our list. Here are the final factorizations:

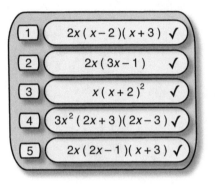

1	$2x(x-2)(x+3)$ ✓
2	$2x(3x-1)$ ✓
3	$x(x+2)^2$ ✓
4	$3x^2(2x+3)(2x-3)$ ✓
5	$2x(2x-1)(x+3)$ ✓

STEP 4: WASH, RINSE, REPEAT.
(FACTOR UNTIL YOU CAN'T FACTOR ANYMORE!)

This last step is really more of a motto to live by when it comes to factoring than a step. Because as you now know, we've already finished factoring our five example polynomials. But the bottom line is this: Don't stop factoring until there's absolutely nothing left to factor. If you do that, you'll always get the right answer . . . guaranteed!

POP QUIZ: FACTOR BY GROUPING

Try factoring the polynomial $x^3 - x^2 - 4x + 4$:

Hint: There's a trick that will come in very handy here. It's called grouping, and the idea is that instead of taking out a common factor from all the terms, try splitting the polynomial into two parts and take out different common factors from each part. In this problem, try finding separate common factors for the first two terms and the last two terms. How far does that get you? Are there any identities that might help here too?

SOLVING QUADRATIC EQUATIONS

To many people quadratic equations are the ultimate symbol of algebra. And rightly so since there are a lot of problems that have quadratic equations in them. In fact, we've seen our fair share of them ourselves—including the challenge problems from the end of chapter 1 that we've been struggling to finish. Until now, that is.

STANDARD FORM OF QUADRATIC EQUATIONS

A quadratic equation is any equation of the form

$$ax^2 + bx + c = 0$$

This is what is called "standard form" for a quadratic equation. Not surprisingly, a, b, and c are constants, and a is not allowed to be zero—because then it wouldn't be a quadratic equation, would it?

✏ POP QUIZ: STANDARD FORM

Put the following quadratic equations into standard form and note the values of the numerical coefficients a, b, and c.

1. $(2x+5)(x-2)=0$

Standard form→ ____ x^2+ ____ $x+$ ____ $=0$

2. $(x-3)^2+1=0$

Standard form→ ____ x^2+ ____ $x+$ ____ $=0$

3. $(x+1)(x-1)+1=0$

Standard form→ ____ x^2+ ____ $x+$ ____ $=0$

Although all quadratic equations can be written in standard form, there are actually three different ways in which you will typically see quadratic equations in problems—and therefore there are three different methods that you'll normally want to use to solve them. The three different variations on the problem can be summarized by the following conditions on the numerical coefficients a, b, and c of the quadratic equation:

- $a \neq 0$, $b = 0$, $c \neq 0 \rightarrow$ no linear term
- $a \neq 0$, $b \neq 0$, $c = 0 \rightarrow$ no constant term
- $a \neq 0$, $b \neq 0$, $c \neq 0 \rightarrow$ all three terms

Let's look at each situation separately.

SOLVING QUADRATICS WITH NO LINEAR TERM

When $b = 0$, the quadratic equation in standard form simplifies to

$$ax^2 + c = 0$$

We've been solving equations like this since the first chapter of the book, so we won't dwell on it too long. The solution can be found by solving for x:

$$x = \pm[-c/a]^{1/2}$$

Remember, these values of x are the roots of the equation. As long as $-c/a \geq 0$ here, then these values of x are also the locations where the polynomial crosses the x-axis. What happens if $-c/a < 0$? Well, then you're taking the square root of a negative number, which means that the roots are going to be imaginary numbers . . . and that the polynomial will not cross the x-axis.

SOLVING QUADRATICS WITH NO CONSTANT TERM

Okay, the second case we'll look at is when the quadratic equation has no constant term. In other words, when $c = 0$. In that case, the standard form quadratic equation simplifies to

$$ax^2 + bx = 0$$

How can we solve this? Well, all we have to do is factor out a common term . . . and that's something we're great at now! The common term that we want to factor here is x, so the equation becomes

$$x(ax + b) = 0$$

This equation has two roots (as all quadratics must). The first is $x = 0$, and the second is $x = -b/a$.

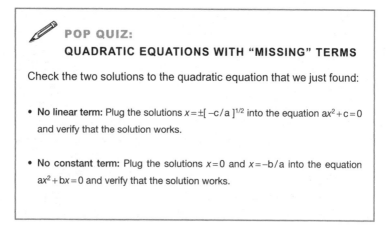

POP QUIZ:

QUADRATIC EQUATIONS WITH "MISSING" TERMS

Check the two solutions to the quadratic equation that we just found:

- **No linear term:** Plug the solutions $x = \pm[\,-c/a\,]^{1/2}$ into the equation $ax^2 + c = 0$ and verify that the solution works.

- **No constant term:** Plug the solutions $x = 0$ and $x = -b/a$ into the equation $ax^2 + bx = 0$ and verify that the solution works.

SOLVING QUADRATICS WITH ALL TERMS

The third and final case to think about for quadratic equations is when all three terms are present:

$$ax^2 + bx + c = 0$$

How do we solve these guys? Well, that all depends on the particular values of the numerical coefficients. We just spent quite a bit of time figuring out one way to solve quadratic equations like this. Yes, that's right—factoring.

WATCH OUT!

Quadratics Must be in Standard Form to Solve by Factoring

To solve a quadratic equation by factoring, you must first make sure that the quadratic is written in standard form—meaning you must make sure that all of the non-zero terms are on the left side of the equation, with only zero on the right side of the equals sign. If you don't do this, then the factors you solve for won't actually be the roots of the equation—and it's much more likely that you'll make an error somewhere.

SOLUTION VIA FACTORING

The equation $x^2 - x - 6 = 0$ is one of an infinite number of quadratic equations that can be solved by factoring. In this case there are no common factors, and this is not a special case where an identity helps us, so we need to factor this quadratic equation using the methods described in step 3 of the factoring tutorial. We know that the equation must factor into something that looks like

$(x - \underline{})(x + \underline{}) = 0$

Since the coefficient on the x^2 term is one, we know that the product of these two roots must equal −6 and their sum must equal −1. If you go through the various possible combinations, you'll eventually arrive at the correct factorization

$(x - 3)(x + 2) = 0$

Next we need to set each of the factors equal to zero to find the roots of the equation: $x - 3 = 0$ and $x + 2 = 0$, which of course give the solutions $x = 3$ and $x = -2$. You can easily check and make sure that these are valid solutions by plugging them back into the equation.

THE QUADRATIC FORMULA

But some equations are difficult (or impossible) to factor this way, and for those equations there is another way around the problem . . . so to speak. It's called, reasonably enough, the quadratic formula:

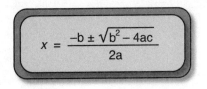

$$x = \frac{-b \pm \sqrt{b^2 - 4ac}}{2a}$$

and it can be used to find the solution of any quadratic equation that's written in standard form.

Where does it come from? Well, one way you can find this general solution to a quadratic equation is by using a very clever technique called "completing the square." But that's a story for another day (although I do encourage you to look it up online if you're interested). The important thing for today is that we use the quadratic formula to solve a quadratic equation by simply writing the equation in standard form, then noting the values of the numerical coefficients, and finally using these values in the formula to find the roots.

WATCH OUT!

Quadratic Formula Pitfalls

1. It's *very* important not to forget about both roots. In other words, don't forget that the ± sign means that this formula gives two values: one in which you add the square root, and another in which you subtract it.

2. If the expression inside the square root evaluates to a negative number, then the quadratic equation you are solving does not have roots that are real numbers. In other words, the roots of such an equation are imaginary. This means that if you run into a problem in which the square root contains a negative number, then that problem has no solution (unless you're using imaginary numbers).

POP QUIZ:

LET'S SOLVE EQUATIONS WITH THE QUADRATIC FORMULA!

Use the quadratic formula to solve the equation $x^2 + 7x - 918 = 0$:

PUTTING A NAME WITH THE FACE . . . AT LAST

It's finally time to take another look at the equation from our challenge problem that we got stuck on way back near the beginning of the chapter:

$$n_1^2 + n_1 - 1056 = 0$$

Lo and behold, it turns out that this nemesis of ours is nothing more than a quadratic equation with coefficients $a = 1$, $b = 1$, and $c = -1056$. And that means that we can solve it using the quadratic formula! Well, that's interesting, don't you think?

SECRET AGENT MATH-LIB

WILL SECRET AGENT MATH ESCAPE THIS NIGHTMARE?

Good evening, secret agent _____ (name of fruit or vegetable). Your mission tonight...um, hello? Secret agent _____ (name of fruit or vegetable)? Are you there?

No, of course you're not.

After decrypting the message "NO FIGHT. RUN!" you did the sensible thing and ran. But it was too late. Just as you moved to make your exit you were grabbed, blindfolded, thrown into the trunk of a car, and driven away. _____ (exclamation)! And when you woke up, you were locked inside the small windowless room in which you're now sitting.

ALGEBRAIC GRAFFITI

After looking around for a minute, you find the pirate carving on the wall saying "YOU DID" that you saw a picture of earlier. With shock, you realize what that means—apparently, you really did carve that message. Which means that you've been in this room before! But you don't remember it. Or…do you? Well, if you were starting to remember it, you're not anymore because you've been distracted by something on the wall next to the pirate carving that appears to be some kind of safe. Here's what you see:

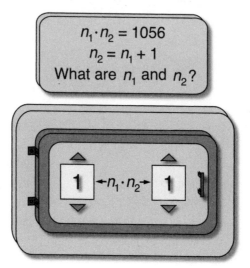

$$n_1 \cdot n_2 = 1056$$
$$n_2 = n_1 + 1$$
What are n_1 and n_2?

My goodness! Little do you—the secret agent "you," that is—know, but this sign seems to be asking the exact same question that you—the reader "you," that is—have been trying to answer all book long! That's right, it's the challenge problem that we got stuck trying to answer at the beginning of the chapter. Translating the sign, you realize that it's asking: If you are told that the product of two consecutive integers is 1056, what are the two integers? And apparently it wants you to use those integers to unlock the safe.

CHALLENGE PROBLEM #2 REDUX

Well, I guess that means it's finally time for us to take that challenge and solve the problem! As you'll recall, at the beginning of this chapter we began solving this problem but got stuck when we needed to solve the equation $n_1^2 + n_1 = 1056$. But now that we know how to solve quadratic equations, it's time for us to get unstuck.

The first step is to get this equation into the right form so that we can solve it using the quadratic formula. In other words, since the quadratic formula is set up to solve equations that look like

$$ax^2 + bx + c = 0$$

we need to move the constant 1056 to the left side of our equation, like this

$$n_1^2 + n_1 - 1056 = 0$$

Then, using the fact that n_1 in our equation serves the same purpose as the variable x in the general quadratic formula, we can determine the values of the numerical coefficients in our particular problem:

$$a = \underline{\quad 1 \quad} \qquad b = \underline{\qquad} \qquad c = \underline{\qquad}$$

As we've learned, the solutions we're after for n_1 are the roots of the quadratic equation with these a, b, and c values. And we can find these roots using:

$$n_1 = \dfrac{-\underline{\quad}_{(b)} \pm \sqrt{\left(\underline{\quad}_{(b)}\right)^2 - 4 \cdot \underline{\quad}_{(a)} \cdot \underline{\quad}_{(c)}}}{2 \cdot \underline{\quad}_{(a)}}$$

After a bit of arithmetic, you should be able to simplify this equation to look like

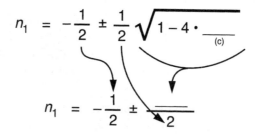

As long as the expression inside that square root symbol evaluates to a positive value, we should be good to go. It does, right? Assuming so, at this point we're ready to find our two solutions. Two? That's right...you were expecting both of them, right? It's a quadratic, so it has to have two roots...and that means that we also have two valid solutions:

solution 1: $n_1 = -\dfrac{1}{2} + \dfrac{\rule{1cm}{0.4pt}}{2} = \dfrac{\rule{1cm}{0.4pt}}{2} = \rule{1cm}{0.4pt}$ _plus!_

solution 2: $n_1 = -\dfrac{1}{2} - \dfrac{\rule{1cm}{0.4pt}}{2} = \dfrac{\rule{1cm}{0.4pt}}{2} = \rule{1cm}{0.4pt}$ _minus!_

So n_1 can be either _____ or _____. Does that mean we're done? Are those the two numbers that we multiply together to get 1056?

Well, no. First of all, the two numbers we're looking for are supposed to be consecutive, and these two numbers definitely are not that. And second, _____ • _____ = −1056 not 1056. Why is this happening? Because these are both values of n_1... which is only half of the problem! The other half of the problem is to find n_2.

So how do we find n_2? Well, look back at the two equations you found written above the safe. And pay extra close attention to the second of the two equations: $n_2 = n_1 + 1$. Do you see what you have to do? For each of the solutions for n_1, we get a corresponding solutions for n_2:

solution 1: $n_2 = \underset{(n_1)}{\underline{\quad}} + 1 = \underline{\qquad}$

solution 2: $n_2 = \underset{(n_1)}{\underline{\quad}} + 1 = \underline{\qquad}$

And that's it! We now have our two complete solutions:

solution 1: $n_1 = \underline{\qquad} \quad n_2 = \underline{\qquad}$

solution 2: $n_1 = \underline{\qquad} \quad n_2 = \underline{\qquad}$

Before moving on, go ahead and try doing the multiplication to make sure everything works. In other words, make sure that $n_1 \cdot n_2$ really does equal 1056 for both solutions.

THE CURIOUS "GIFT"

Okay, it's time for secret agent math to enter this solution into the lock on the safe and see if we can open up that door. But which solution should we use? In principle it shouldn't matter since they're both perfectly valid. But then you look again at the door and see a tiny note:

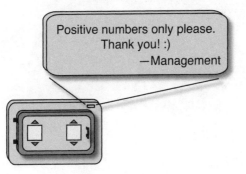

Well, that certainly makes the decision easier. So, armed with this new information, you set the left number on the safe door to the value of n_1 and the right number to the value of n_2. And, presuming you entered 32 on the left and 33 on the right, the door swings open. If those weren't the numbers you entered, the door remained locked, an annoying buzzer sounded, and a voice suggested you go back and try again before moving on.

Okay, you've got the door open. So what's inside? This...

What?! A plant? This is getting bizarre. First the inexplicably polite and even enthusiastic message left by your captors letting you know to use the positive solution to open the safe. And now they've given you a plant! But then you notice a message written on the bottom of the pot...

Um…that can't be good. And all at once you realize that not only are there no windows in this room, but there aren't even any doors! Not bothering to dwell on how you could have gotten into the room in the first place, you make a terrifying connection: Your air supply is limited, there's nothing getting in or out of this room, and somebody thought it was very funny to give you a spindly four-leafed plant that probably couldn't make enough oxygen to keep an ant alive. So what does a math secret agent like you do in a situation like this? Do you panic and cry?

CAN MATH SAVE YOUR LIFE?

No, you plan to run the numbers and estimate how much time you have before you run out of air. But there's a problem: you've never solved this problem before…they didn't teach you how to do it in secret agent math academy. So are you out of luck? Again, no. It's definitely a little intimidating (especially since you're worried about your well-being), but a math secret agent's rule of thumb when working out problems like this is to take it slow and start with what you know.

So, you figure that if you knew how big the room is, and how big one of your breaths is, then you'd be able to figure out how many total breaths of air the room contains. And then if you could figure out how quickly you breathe—say, how many breaths you take per minute—then you could figure out how quickly you're going to deplete your air supply. And if you know that, then you know how much time you have…to escape! So, in a way, math just might save your life.

USING MATH TO MAKE ESTIMATES

You start by figuring out how big the room is. You take a look around and conclude that it's pretty close to being shaped like a cube. Which means that its height, width, and length are all approximately the same size. But exactly how big? Well, you don't actually need to know *exactly* how big it is—a reasonable estimate will go a long ways.

You know that your typical stride is about 1 meter long (a little more than 3 feet). So you pace off the width of the room and find that it's about 3 m wide (meters is abbreviated "m"). And since it's almost a cube, you figure that it's also about 3 m long and 3 m high. The best way to put this all together to measure the total size of the room is to find its volume: $3\,m \cdot 3\,m \cdot 3\,m = (3\,m)^3 = 27\ m^3$. Remember, the "m" here isn't a variable, it's here to let you know precisely what units we're talking about—in this case meters.

Now what do you do? Well, that was the easy part. Next you need to figure out how many total breaths of air the room contains. But the problem is you don't have a clue how "big" your breaths are…or do you? At some point in your life, you remember exhaling into a plastic bag with the intent of popping it to freak out the cat. And you realize this means that the volume of one of your breaths is about the same as the volume of that bag. Now, you don't know *exactly* how big that bag was, but that doesn't matter—you just need an estimate. So, you estimate that one of your breaths would fill a bag that's at most about 10 centimeters (or almost 4 inches) on a side. Maybe a little more, maybe a little less, but that's a good enough number to get you started.

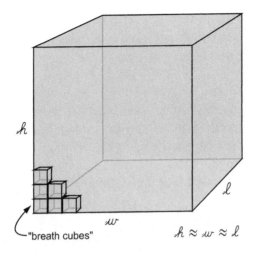

"breath cubes" $\hbar \approx w \approx \ell$

It turns out that 10 cm (centimeters is abbreviated "cm") is equal to 0.1 meters (since 1 m = 100 cm), which means that the volume of one of your breaths—we'll call it a "breath cube"—is about 0.1 m • 0.1 m • 0.1 m = $(0.1 \, \text{m})^3 = 0.001 \, \text{m}^3$.

Wow! That doesn't seem like much, does it? Well, let's see. Let's calculate how many breaths of air the room contains. All we need to do to get that number is to divide the total volume of the room by the volume of one of your breath cubes. And that's just

number of breaths in room $\approx 27 \, \text{m}^3 / 0.001 \, \text{m}^3 = 27{,}000$

Well, that seems like a pretty big number. Maybe you're going to be okay after all. But then again, you do breathe all the time...like constantly. So maybe that not really such a big number. There's no sense in wondering when you can do the calculation and see.

CONVERTING UNITS LIKE YOUR LIFE DEPENDS ON IT

So the question now is how long will 27,000 breaths last. An hour? A day? A week? All we need to know to find out is an estimate of how frequently you breathe. In other words, if you breathe one time per second, then you would have 27,000 seconds worth of air in the room. But unless you've been sprinting, you probably don't breathe that fast.

Instead of guessing, you decide to do the experiment and count how many breaths you take in one minute. For most people that number is between 15 and 20 breaths per minute. And since you're currently under a bit of stress trying to orchestrate your escape, you find that you're breathing at a rate of about twenty breaths per minute. Which means that the 27,000 breaths in the room will last

27,000 breaths / 20 breaths per minute = 1,350 minutes

How long is that? Well, since there are sixty minutes in an hour and twenty-four hours in a day, you calculate that those 27,000 breaths will last you

$$1350 \text{ minutes} \cdot \frac{1 \text{ hour}}{60 \text{ minutes}} \cdot \frac{1 \text{ day}}{24 \text{ hours}} \approx 1 \text{ day}$$

And that's the answer—you've got a little less than one day to figure out how to get out of there. In truth, the news could have been better…but you figure that it could have been a lot worse too.

Besides, you also know that the estimate you've done is the absolute worst-case scenario. For one thing, your breathing rate will probably settle a bit once you calm down, and that alone will buy you some more time. But the real saving grace is that each breath you take doesn't completely deplete the oxygen in the air you inhale. You only use a fraction of the oxygen in each chunk of air every time that you breathe it. Which means that each of those breaths in the room is actually good for several breaths' worth of oxygen. So you figure you've probably got days before you need to make your escape.

GOOD MORNING!

Feeling calm and confident that you'll figure something out tomorrow, or at the worst the next day, you decide to stop worrying and get some rest. You curl up on the floor, close your eyes, and go to sleep...only to be awoken moments later by a horrendous clatter and the obnoxious sound of...the trash truck. You look around the room and realize that it's not sealed, it's not windowless or doorless, and you're not a math secret agent. But you are tired. So you roll over and go back to sleep.

Does secret agent math escape from this nightmare? Yes, literally! So long, secret agent _____ *(name of fruit or vegetable)*. Sweet dreams!

- -

OVERACHIEVER BADGE

HOW TO USE PROPORTIONS TO CONVERT UNITS

One simple but mighty handy thing that algebra can help you with is converting from one system of units to another. In fact, we just saw this in action when secret agent math converted from some number of minutes into some number of days. There are endless uses for this type of conversion. For example, imagine you're given some distance in miles, and you'd like to know what that distance is in kilometers, inches, fathoms, light-years, or whatever. Well, algebra, and specifically "proportions," can help with that.

　　To demonstrate the method, let's look at the specific example of converting a distance of twenty-six miles (the

Cont'd

Cont'd from page 399

approximate length of a marathon) into kilometers. Here's
how it works:

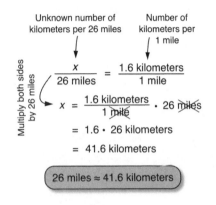

The equation on top is what's called a proportion. What's
that? Well, a proportion is just a specific type of equation that
relates two ratios. In this case, the left side of the equation is
the ratio of some unknown number of kilometers to some
known number of miles. Similarly, the right side of the equation is the ratio of some known number of kilometers to
some known number of miles. In other words, both sides relate the same type of one thing in their numerators to the
same type of another thing in their denominators.

And since three of the four parts of the equation are known,
we can use some simple algebra to solve for the fourth and
final component, *x*. Where did the 1.6 kilometers per mile
come from on the right side of the equation? Well, that's just
a standard conversion that you can easily find in books or on
the Internet. If you had wanted to convert from miles to light-
years instead, the only difference is that you'd use a ratio saying that one light-year is 5.9 trillion miles on the right side of
the equation.

CHALLENGE PROBLEM #3

"LOOK OUT BELOW!"

At long last it's time to solve the third and final challenge problem from the first chapter. In case you can't remember what that question was, here's a refresher:

> *You drop a rock into a well and hear a splash two seconds later. How deep is your well?*

While it's unlikely that you'll ever encounter this exact problem in the real world (there aren't a lot of open wells around anymore), the problem has many features in common with a ton of other real-world problems from math and physics that you will likely run into someday.

But before we get started, let me just say that there are a lot of steps to this problem. And that's because this really is a real-world problem. In case you haven't noticed yet, problems in the real world don't usually look like problems in textbooks. Real-world problems have multiple parts, possible ambiguities that you have to figure out, and the like. They're much more complicated to solve, but they're much more rewarding to solve too. So, let's take a look, shall we?

MAKE A PLAN TO SOLVE THE PROBLEM

The first thing you should do when you start working on a problem is make sure you understand what you're actually trying to solve for. In this case, it's not entirely clear what is meant by "how deep is your well." Are we talking about how deep to the bottom of the well way down beneath the water? Or are we talking about how far it is from the top of the well to wherever the water that the rock splashed in begins? In this case, we're looking for the second of these two things. So now that we're sure what we're looking for, the next thing to do is think about how we should go about solving the problem.

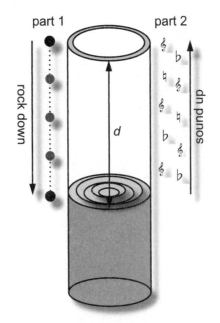

As you can see in this drawing, we can break the problem up into two parts. The first thing that happens is that you drop the rock and let it fall to the water a distance that we'll call "*d*" below the point where you're standing. This fall takes a certain amount of time which we'll call t_{down}. We're told in the problem that it takes a total of two seconds from the time you drop the rock to the time you hear the splash, so we know for sure that $t_{down} <$ two seconds. The second part of the problem starts as soon as the rock hits the water and causes the splash that you eventually hear at the surface a time t_{up} later. So to recap:

- t_{down} is the time it takes the rock to fall from your hand down to the surface of the water and make a splash
- t_{up} is the time it takes the sound made by this splash to travel from the water and up to your ear

So that's the setup. Now let's look at these two parts of the problem separately and figure out how to write equations that will allow us to solve for the two times t_{down} and t_{up}.

PART ONE: THE WAY DOWN

Up first, let's look at the rock's journey from your hand down to the surface of the water. Before we do anything, we need to talk about a little physics. In particular, we need to figure out how to use physics to write an equation that tells us how far the rock has fallen a certain amount of time after leaving your hand. And since this isn't a physics book, I'm just going to tell you the answer. If you drop a rock (drop, not throw) from an initial height d over the surface of the water and then let it fall, the height of the rock above the water a time t later is given by the function

$$h(t) = -\tfrac{1}{2}g\,t^2 + d$$

In this quadratic binomial equation, g is called the "gravitational acceleration," and its value influences how fast a falling object accelerates. The value of g is different on different planets, but on earth (where I'm assuming our story is taking place), g has an approximate value of $9.8\,\text{m/s}^2$ (those units are "meters per second squared" . . . but you don't need to worry about that).

What we're really after here is figuring out how long it takes the rock to fall and hit the water. And at this point, we actually know enough to do that. When the rock hits the water after a time which we've called t_{down}, its height above the water is zero. We can write this statement in the form of this very simple equation:

$$h(t_{\text{down}}) = 0$$

Earlier we came up with an equation for $h(t)$, and if we plug the time t_{down} into that equation, we get

$$h(t_{\text{down}}) = -\tfrac{1}{2}g\,t^2_{\text{down}} + d$$

But we also just figured out that $h(t_{\text{down}}) = 0$, so that means we can combine these two things to get

$$-\tfrac{1}{2}g\,t^2_{\text{down}} + d = 0$$

Okay, what does this mean? Well, it's an equation that gives a relationship between the time it takes the rock to fall, t_{down}, and the distance it falls, d. We can make this clearer by doing a bit of algebraic manipulation and solving for d. All we have to do is move the $-\frac{1}{2}g\,t^2_{down}$ from the left to the right side of the equation, and we get

$$d = \frac{1}{2}g\,t^2_{down}$$

This says that the distance d depends on the second power of the time. In other words, this is a quadratic equation. And if you think about it, you'll see that this means that if the amount of time a rock takes to drop is twice as long in one well as another, then that well with the longer drop time must be $2^2 = 4$ times deeper.

PART TWO: THE WAY BACK UP

We now have an equation that describes the relationship between the variables in the first part of the problem, which means that it's now time to move on to the second part of the problem dealing with the time it takes the sound from the splash to reach your ear. In this part we need to think about another idea from physics: speed. In particular, we need to figure out what speed is and how we can calculate it. As you probably know, the speed of an object is just how fast it's moving. And by how "fast," I mean how far it moves in a certain amount of time. Which means that the speed of an object is just the distance it travels divided by the time it takes to cover that distance.

Okay, but what does that have to do with the second part of the problem? Well, when the rock hits the water it causes a splash. And that splash is accompanied by a sound that travels up the well to your ear at a certain *speed*. Aha! The distance from the surface of the water to your eagerly anticipating ear must equal the speed at which sound travels times the time it takes it to travel to you.

In other words, if we call the speed of sound in air v (which is about 353 m/s), then we can write an equation that relates the depth of the well and the time it takes sound to traverse that depth, t_{up}, like this

$$d = v \bullet t_{up}$$

And now, just as we have an equation relating the variables from the first part of our problem, we also have an equation relating the variables from the second part. Which means that we're in business.

HALFTIME RECAP

Let's take stock of what we have. So far we've put together equations that describe the relationship between distance and time for both parts of our problem:

- Part One → $d = \frac{1}{2}g\, t^2{}_{down}$
- Part Two → $d = v \bullet t_{up}$

That's great, but there's one little problem. We've got two equations here, but three unknowns: t_{down}, t_{up}, and d (remember that $g \approx 10\,m/s^2$ and $v \approx 353\,m/s$ are both constants, not variables). As we now know, solving for three unknowns requires three equations.

THE ELUSIVE THIRD EQUATION

Does that mean we're out of luck? No, there's actually another equation staring right at us that we haven't bothered to write down yet. And when we do, it'll give us that all important third relationship that will allow us to solve the problem. Can you figure of what that equation is? Here's a hint: It has something to do with the relationship between the two times t_{down} and t_{up}, and the total time between the moment you drop the rock to the moment you hear the splash. As you'll recall, we were told in the problem that this total time is two seconds, which means that

$$t_{down} + t_{up} = 2$$

This just says that the total time for the two parts of the problem must equal the total time for the entire problem! It's pretty simple, but it's very useful because it allows us to eliminate either t_{down} or t_{up} from our equations. And when we do that, we'll be left with a system of two equations for our two unknown values—and that's something we've had a lot of practice solving!

FROM THREE UNKNOWNS DOWN TO TWO

So let's go ahead and eliminate one of the variables. Which one? Well, you could eliminate either t_{down} or t_{up}, but the algebra works out a little easier if we get rid of t_{up}, so that's what we'll do. First, solve $t_{down} + t_{up} = 2$ for t_{up} like this:

$$t_{up} = 2 - t_{down}$$

Then plug this into the equation we found in the second part of the problem, $d = v \bullet t_{up}$, like this:

$$d = v \bullet (2 - t_{down})$$

And that does it—t_{up} is out of the picture . . . for now. What should we do next? Well, let's take a look at what we're dealing with. We now have two equations for the two unknown values, d and t_{down}. They are:

- Our original equation relating d to t_{down}: $d = \frac{1}{2} g\, t^2_{down}$
- Our new equation also relating d to t_{down}:
  $d = v \bullet (2 - t_{down}) = 2v - vt_{down}$

IT'S A QUADRATIC EQUATION!

All right, let's now solve this system of two equations. In this case, since they're both equal to d, all we have to do is set them equal to each other. When we do that we get

$$\frac{1}{2} g\, t^2_{down} = 2v - vt_{down}$$

Is this starting to look familiar? How about if we move everything that's on the right side of the equation over to the left, like this:

$$\tfrac{1}{2}g\, t^2{}_{\text{down}} + v t_{\text{down}} - 2v = 0$$

Well, would you look at that—it's a quadratic equation for the variable t_{down}! And that's something we're quite capable of solving.

Comparing our equation to the standard form of a quadratic equation, $ax^2 + bx + c = 0$, we see that our quadratic equation has numerical coefficients equal to $a = \tfrac{1}{2}g$, $b = v$, and $c = -2v$. Using these values, we can use the quadratic formula to find the roots of the equation—and the solutions for t_{down}. Here's what we get:

$$t_{\text{down}} = \frac{-v \pm \sqrt{v^2 - 4 \cdot \tfrac{1}{2}g \cdot (-2v)}}{2 \cdot \tfrac{1}{2}g}$$

$$= \frac{-v \pm \sqrt{v^2 + 4gv}}{g}$$

We can now break this apart into our two solutions. As always, the first has a plus sign, and the second has a minus sign:

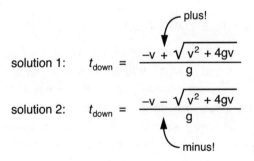

solution 1: $t_{\text{down}} = \dfrac{-v + \sqrt{v^2 + 4gv}}{g}$

solution 2: $t_{\text{down}} = \dfrac{-v - \sqrt{v^2 + 4gv}}{g}$

Finally, we're ready to plug in some numbers to calculate the numerical values of t_{down}:

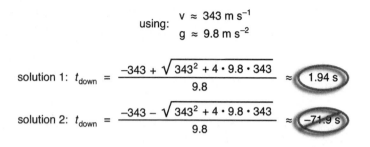

As you can see, the first solution is $t_{down} \approx 1.94$ seconds. You can also see that I've crossed out the second solution, which is approximately –71.9 seconds. You should *not* take this to mean that –71.9 is an invalid or incorrect solution to the equation. In fact, it's a perfectly valid solution to the math problem. But –71.9 seconds is *not* a valid solution to the physics problem since it doesn't make any sense to say that it takes the rock about –71.9 seconds to reach the water. So, in this case the solution we're after is the first one—about 1.94 seconds.

Let's think a little bit about what this answer means. We know that it takes a total of two seconds for the rock to fall from your hand, make a splash, and for you to hear the sound caused by that splash. But we now know that it takes about 1.94 seconds just for the rock to reach the water. Which means that it takes sound only $2 - 1.94 \approx 0.06$ seconds to reach your ear! Sound is fast!

HOW DEEP IS YOUR WELL?

While that speed is quite impressive, we haven't actually finished the problem yet. We haven't? No, the original question asked us to figure out how deep the well is, not how long it takes the rock to traverse this distance. So, the final thing we need to do is figure out how to use what we've found out so far to answer this last question. And fortunately, that's not too difficult.

If you look way back near the beginning of the problem, you'll see that we wrote down an equation that related the depth of the well, *d*, to the time it takes the rock to fall down it, t_{down}:

$$d = \tfrac{1}{2} g\, t^2_{\text{down}}$$

And that equation is exactly what we need. We know the value of g, and now we also know the value of t_{down}. Which means that we can use this equation to calculate d, like this:

$$d = \tfrac{1}{2} \cdot 9.8 \cdot 1.94^2 \approx 18.4\,\text{meters}$$

That's a pretty deep well! For those of you more comfortable with feet and miles, 18.4 meters is about 60.4 feet—so that rock had itself quite an adventure. And, I think you'll agree, we did too.

GAME OVER?

Game over? No way. There's always more math to learn about and explore. Book over? Yes, definitely. And it's been an absolute pleasure having you along for this journey. I hope that I've been able to help you see and understand why I claim that the world of algebra is a captivating place to explore. It's okay if you don't absolutely love it . . . not everybody does, and not everybody will. But I do hope that you at least come away with an appreciation for why many people do.

Looking at math is a lot like looking at a tree. From a distance, a tree looks like nothing more than a giant mess of individual leaves. But as you get closer and begin to look at the details of the tree, you see that all those leaves are actually connected . . . that they're tied together by branches. First small ones, and then increasingly larger as they move down toward the trunk. From a distance, math too can look like nothing more than a giant heap of disconnected pieces—like a pile of leaves. But as you start to investigate what's hidden beneath those mathematical leaves, you uncover a beautiful and intricate web of ideas that tie it all together.

Yes, algebra can be serious—especially when a grade, SAT score, or business deal are on the line. But at its heart, algebra . . .

and math . . . are inventions of the human mind. And they're here to entertain us.

It's free, it's fun, it's fascinating, it's algebra.

· FINAL EXAM ·

• ROOTS OF POLYNOMIALS

1. Use the fact that any polynomial can be written as the product of a constant and a number of linear terms containing its roots, like this

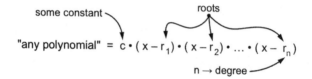

$$\text{"any polynomial"} = c \cdot (x - r_1) \cdot (x - r_2) \cdot \ldots \cdot (x - r_n)$$

to find equations for the two polynomial shown on this graph:

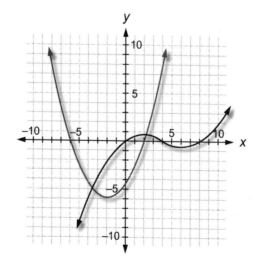

You'll need to find the roots of each polynomial, write an equation like the one given (ignoring the constant for now), and then use FOIL to multiply everything out. After you have your polynomial, calculate its value at a few x positions and see if they match the graph. If not, think about whether or not including a constant, c, might help.

- **FACTORING POLYNOMIALS**

2. In the Math Brain Game, we looked at the geometrical meaning of two algebraic identities. Now, here's your chance to take a look at the meaning of a third of the identities we saw in that section. Use FOIL to multiply out the left side of this identity and interpret the drawing below.

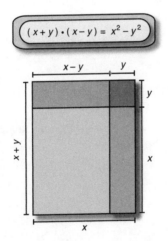

$$(x+y) \cdot (x-y) = x^2 - y^2$$

- **SOLVING QUADRATIC EQUATIONS**

3. While this isn't directly about solving a quadratic equation, this question is at least related to a problem in which we solved one . . . so I figure it's fair game. And it's the last question of the

whole book, so why not do something crazy, right? Here's the question . . .

During the problem about the rock falling into the well, we came up with this equation

$$d = \frac{1}{2} g \, t^2_{\text{down}}$$

that tells us how deep the well is depending on how long it took the rock to fall. But we can turn this equation around and solve for how long it takes the rock to fall for a given distance too, like this:

$$t_{\text{down}} = \sqrt{2d/g}$$

So the deeper the well, the longer it takes the rock to hit the bottom. But there's another parameter in that equation—the little g, which has to do with how strongly the earth's gravity is pulling on and accelerating the rock. So the question is this: Is the value of g bigger or smaller on the moon than it is on earth?

Think about videos you've seen of people walking on the moon—everything seems like it's in slow motion, right? Which would mean that it must take the rock longer to fall on the moon. But what does that say about g?

THE MATH DUDE'S
SOLUTIONS

———

PROLOGUE

Pop Quiz: *Name that Number Property*   (page 11)

	1	99	−5	6	0	24	−8	−1
Am I even (E) or odd (O)?	O	O	O	E	E	E	E	O
Am I positive (P) or negative (N)?	P	P	N	P	*	P	N	N

* The "trick" question was whether zero is a positive or negative number. The answer is that **zero is neither positive nor negative!**

Secret Agent (in Training) Math-Lib: *Is Math Boring?*   (page 14)

(A)   4 → "four"

(B)   7 → rhymes with "eleven"

(C)   11 → 10, 11, 12, 13, 14, 15, 16, 17, 18, 19 (11 has two 1s!)

(D)   3 → 3 and 7 both have two of the letter "e", but 7 has already been used

Once you've got these numbers, you should be able to use the following decoded secret message answering the age-old question "Is math boring?" to figure out the answers to the arithmetic problems. Here's the decoded message:

N   O   P   E !

Keep an O   P   E   N mind!

Pop Quiz: *It's Hip to Be a Perfect Square*   (page 20)

1.   55 → Am I a perfect square? _N_ If yes, then I'm equal to ___ • ___
2.   100 → Am I a perfect square? _Y_ If yes, then I'm equal to **10** • **10**
3.   81 → Am I a perfect square? _Y_ If yes, then I'm equal to **9** • **9**
4.   68 → Am I a perfect square? _N_ If yes, then I'm equal to ___ • ___
5.   121 → Am I a perfect square? _Y_ If yes, then I'm equal to **11** • **11**

Bonus question: $2^6 = 64$. Yes, 64 is a perfect square → 8 • 8 = 64.

Final Exam

• **Number Properties**

These solutions are just examples. You can use whatever even and odd numbers you like.

1. Think of an even number: **4** . Now think of another even number: **8** .
   - Add them together: **12** .
   - Is the answer even or odd? **E**
   - Subtract the first from the second: **4** .
   - Is the answer positive or negative? **P**
   - Is the answer even or odd? **E**
   - Multiply them together **32** .
   - Is the answer even or odd? **E**

2. Think of an odd number: **7** . Now think of another odd number: **3** .
   - Add them together: **10** .
   - Is the answer even or odd? **E**
   - Subtract the first from the second: **–4** .
   - Is the answer positive or negative? **N**
   - Is the answer even or odd? **E**
   - Multiply them together **21** .
   - Is the answer even or odd? **O**

3. Think of an even number: **2** . Now think of an odd number: **5** .
   - Add them together: **7** .
   - Is the answer even or odd? **O**
   - Subtract the first from the second: **3** .
   - Is the answer positive or negative? **P**
   - Is the answer even or odd? **O**
   - Multiply them together **10** .
   - Is the answer even or odd? **E**

Do you notice any patterns here? Here's a quick and dirty summary of the patterns that have emerged in these problems:

	even even	odd odd	even odd
add subtract	even	even	odd
multiply divide	even	odd	even

In other words, adding or subtracting a pair of even numbers gives an even result, multiplying or dividing a pair of odd numbers gives an odd result, and so on.

• Notation for Arithmetic

4. "Thirty-two plus twenty-eight minus fifty-nine." $32 + 28 - 59$

5. "Eighteen times two divided by nine times two." $18 \cdot 2 / 9 \cdot 2$

6. "Eighteen times two, all divided by nine times two." $(18 \cdot 2) / (9 \cdot 2)$

Beware! English to math translations can sometimes be ambiguous. For example, problem 6 could have also been answered $(18 \cdot 2)/9 \cdot 2$.

• Squaring Numbers

This is just an example. You could have used any number you like.

7. Pick a number between 1 and 10: __5__ . Let's call this number a.
   • Square that number → $a^2 =$ __25__ . Let's call this number b.
   • Multiply your original number by 2 → $2 \cdot a =$ __10__ . Let's call this number c.
   • Subtract c from b → $b - c =$ __15__ . Let's call this number d.
   • Add 1 to d → $d + 1 =$ __16__ . Let's call this number e.

8. Now start again with the same number, a, that you picked before.
   • Subtract 1 from that number → $a - 1 =$ __4__ . Let's call this number f.
   • Square this number → $f^2 =$ __16__ . Let's call this number g.

Surprise! e and g are the same!

• Searching for Patterns

9. 1, 2, 4, 8, 16, 32, . . .   What's the next number? __64__

(Each number is equal to the previous number times 2.)

10. 2, 5, 8, 11, 14, 17, . . .   What's the next number? __20__

(Each number is equal to the previous number plus 3.)

11. 1, 3, 7, 15, 31, 63, . . .   What's the next number? __127__

(Each number is equal to twice the previous number plus 1.)

12. 1, 1, 2, 3, 5, 8, 13, . . .   What's the next number? __21__

(Each number is equal to the sum of the two previous numbers. This is the famous Fibonacci sequence.)

CHAPTER 1: TAKING ALGEBRA TO THE STREETS

Pop Quiz: *Which is Bigger? Smaller? Equal?*   (page 38)

1. The size of the moon is __>__ the size of a mountain
2. The numbers 5, 8, 13, and 21 are __≥__ the number 5.
3. 100 __=__ 10^2
4. The size of the planet Saturn is __<__ the size of the planet Jupiter.
5. The numbers 5, 8, 13, and 21 are __≤__ the number 21.

Pop Quiz: *Can You Use Algebra to Make a Budget?*   (page 41)

Just as we did for our grocery shopping problem, let's create a bunch of variables that we'll call b_1, b_2, b_3, and so on, to represent the cost of each of our monthly bills. Perhaps they actually look something like:

$b_1 \rightarrow$ rent, $b_2 \rightarrow$ groceries, $b_3 \rightarrow$ electricity, etc.

But for now, we don't actually care what they are . . . because we're only interested in coming up with an inequality that we can use every month to figure out if we're going to make our budget. Okay, let's now create a variable called p, which represents the size of your monthly paycheck. Now all we have to do is write an inequality that tells you whether or not your bills are less than or equal to your paycheck.

$$b_1 + b_2 + b_3 + \ldots \leq p$$

If that inequality is true, then you're good to go!

Pop Quiz: *How Many Right Angles Can You Find?*   (page 44)

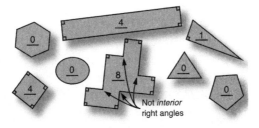

Not *interior* right angles

Pop Quiz: *Calculating Square Roots*   (page 49)

1. $\sqrt{16} = \underline{4}$
   - this is solved **exactly**
   - check your answer: $4^2 = 4 \cdot 4 = 16$ ✓
   - remember that $\sqrt{16}$ is not equal to –4!

2. $x^2 = 1001$. So, $x \approx$ **between 31 and 32 or between –31 and –32**
   - this is solved by **approximation**
   - a more accurate calculator approximation is 31.64 *or* –31.64
   - check your answer: $31.64^2 \approx 1001.1$ (pretty close) ✓
   - check your answer: $(-31.64)^2 = -31.64 \cdot -31.64 \approx 1001.1$ ✓

3. $\sqrt{-1} = \underline{\textbf{undefined}}$
   - we didn't solve this one since we can't take the square root of a negative number!

4. $x^2 = 81$. So, $x = \underline{\textbf{9 or –9}}$
   - this is solved **exactly**
   - check your answer: $9^2 = 9 \cdot 9 = 81$ ✓
   - check your answer: $(-9)^2 = -9 \cdot -9 = 81$ ✓

5. $\sqrt{3.14159} \approx$ __1.77__
   - this is solved with a **calculator** (but it's still not exact)
   - check your answer: $1.77^2 \approx 3.13$ (pretty close) ✓
   - remember that $\sqrt{3.14159}$ is not equal to -1.77!

6. $x^2 = -42$. So, $x =$ **undefined**
   - we didn't solve this one at all because we can't take the square root of a negative number.

7. $\sqrt{55} \approx$ __between 7 and 8__
   - this is solved by **approximation**
   - a more accurate calculator approximation is 7.42
   - check your answer: $7.42^2 \approx 55.1$ (pretty close) ✓
   - remember that $\sqrt{55}$ is not equal to between -7 and -8!

8. $x^2 = 111$. So, $x \approx$ __10.54 _or_ –10.54__
   - this is solved with a **calculator** (but it's still not exact)
   - check your answer: $10.54^2 \approx 111.1$ (pretty close) ✓
   - check your answer: $(-10.54)^2 = -10.54 \bullet -10.54 \approx 111.1$ ✓

Secret Agent (In Training) Math-Lib: *What Is Pythagoras' Message?*   (page 53)

Pythagoras' secret message for math secret agents is . . .

To be a math __S__ __P__ __Y__ ,
you must take an __O__ __A__ __T__ __H__ to occasionally talk like a pirate.
__A__ __R__ __G__ __H__ !

By the way, it's been reported that some pirates prefer to say "Arrr" instead of "Argh." But I have it on good authority that Pythagoras preferred the latter . . . so that's what we're going with.

Algebra Tutorial: *How to Make a Graph*   (page 63)

Step 2

To get each of the c values, plug the value of a into the equation $c = \sqrt{2} \cdot a$. You should be able to figure out values of a from the values of c by looking at the pattern that develops. The (*a*, *c*) ordered pairs on the right are made from the *a* and *c* values on the left.

Step 3

The completed (*x*, *y*) ordered pairs of points B and C are:

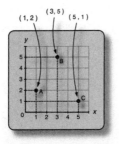

Here are points from the table above plotted on our graph. You'll need to use a calculator to figure out the approximate values of $10\sqrt{2}$, $20\sqrt{2}$, etc., so that you can plot them on the graph.

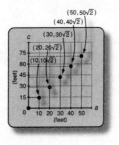

Step 4

You can find the diagonal lengths, *c*, for *a* = 15 and *a* = 35 feet by starting at those locations on the *a*-axis, moving up to the diagonal line, and then moving left to the *c*-axis. The exact values are given by $c = \sqrt{2} \cdot a$.

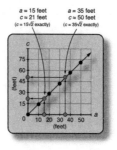

Pop Quiz: *Connect the Dots*   (page 70)

Notice that the scales on the *x* and *y* axes are different. In other words, the *x*-axis goes from 0 to 10 and the *y*-axis goes from 0 to 100. If we didn't do that, we wouldn't be able to see most of the points . . . since they'd be off the top of the plot.

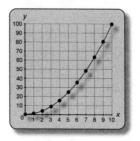

Bonus question: The parts of the curve between the points representing the perfect squares are all the locations of the squares of non-whole numbers (which we'll talk a lot more about in chapter 3). In other words, the squares of all the fractional numbers like 1/2, 51/4, and so on, are represented by the curve.

Final Exam   (page 72)

• **Algebra in the Everyday World**

1. There's no right or wrong answer here. As long as you thought about it a bit, you get 100 percent credit. Wouldn't it be awesome if all tests in life were graded so easily?

• Inequalities

2. Where more than one correct answer is possible, all solutions are shown:
   - three perfect squares past 1 <u>< or ≤</u> $8/(6+2) \cdot 17$
   - $(2+8) \cdot 10$ <u>> or ≥</u> $(2+8)/10$
   - $2+5$ <u>=, ≤, or ≥</u> $5+2$
   - $(8+2)/10$ <u>< or ≤</u> $(8+2) \cdot 10$
   - $(9 \cdot 2) \cdot 8/(6+2)$ <u>> or ≥</u> three perfect squares before 49

• Square Roots

3. You could say that the square root of the square shown in the problem is **4** since that's the square root of the total number of little squares it contains (also known as its area). But what is 4 in that square? It's the length of one of the sides of the square. Which means you could say that the square root of a square is the length of one of its sides. If we're a little more general and instead talk about a square that has sides of some unknown length x, then the square has a total of x^2 little squares. And the square root of that square—that is, the square root of the total number of little squares contained in that square—is just x. Make sense?

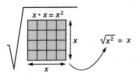

• The Pythagorean Theorem

4. In this problem we know that $a=3$ feet and $b=9$ feet, and we need to use the Pythagorean theorem to figure out the value of c—the height of the ladder. So, $a^2+b^2=3^2+9^2=9+81=90$. Which means that $c^2=90$, and therefore $c=\sqrt{90} \approx 9.5$ feet. Therefore, at a minimum you need a 9.5-foot-tall ladder. In reality—which, after all, is where we live—you probably want a slightly taller ladder to make sure you have a little (literal) wiggle room.

5. In this problem, we know that $c=12$ feet. We also know that the distance from the bottom of the ladder to the house, a, should be one-third the height of the ladder. In other words, $a=\frac{1}{3} \cdot c=\frac{1}{3} \cdot 12=4$ feet. Now we just need to solve for the height of the roof, b, using the Pythagorean theorem: $a^2+b^2=c^2$. If we subtract a^2 from both sides of this equation (we'll have a lot more to say about doing this type of stuff in the

next chapter), we get $b^2 = c^2 - a^2$. Now, if we take the square root of both sides, we get

$$b = \sqrt{c^2 - a^2} = \sqrt{12^2 - 4^2} = \sqrt{144 - 16} = \sqrt{128} \approx 11.3 \text{ feet}$$

- **Graphing Points and Lines**

6. Both of the relationships produce lines. The line $y_2 = 2 \bullet x$ increases much more rapidly than the line y_1.

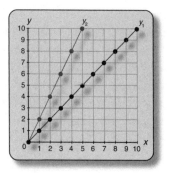

CHAPTER 2: ALGEBRA BASICS

Pop Quiz: *The Advantage of Using Dots?*   (page 79)

The main advantage that dots have over numerals is that they give you a way to visually understand what you're doing. As a simple example of this that's similar to what we found in the "positively odd" Brain Game from the prologue, numerals don't provide much insight about the connections between arithmetic problems like $3 + 3 + 3 = 9$ and $3^2 = 3 \bullet 3 = 9$. Why are the answers the same? It's not immediately obvious with numerals (although once you know the trick it seems like it must be obvious), but with a picture using boxes or dots, it's a lot clearer how the problems are related . . . and why their answers are the same:

$$3 + 3 + 3 = 9$$

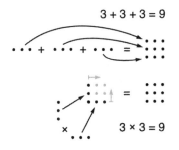

$$3 \times 3 = 9$$

Of course, for just about every other reason, using numerals for arithmetic is a lot easier!

Pop Quiz: *What's the Best Variable Name?*   (page 83)

1. A good name for the length of your new sailboat is the letter L.
2. A good name for the height of the ceiling in your garage is the letter h or H.
3. A good name for the weight of the boat is the letter W.
4. A good name for the speed of your sailboat is s_{boat}. (Why the subscript? Check out the next answer.)
5. A good name for the speed of the wind is s_{wind}. Notice that I used the same variable name, s, for both speeds, and used a subscript to describe what each speed refers to.

Pop Quiz: *English to Expression Translation*   (page 87)

As discussed in the Pop Quiz, all of the problems are ambiguous. Here are some of the possible answers for each question:

1. "three times □ squared plus □ minus ten"
$$3\,\square^2 + \square - 10 \quad or \quad (3 \bullet \square)^2 + \square - 10$$
2. "square root of five times □ divided by □"
$$\sqrt{5} \bullet \square/\square \quad or \quad \sqrt{5 \bullet \square}/\square$$
3. "two plus eight times six divided by three"
$$2 + 8 \bullet 6/3 \quad or \quad (2+8) \bullet (6/3) \quad or \quad (2 + 8 \bullet 6)/3,$$
and so on.

Algebra Tutorial: *The Order of Operations*   (page 88)

3. $5 + 4 \bullet 3/2 - 1^2 = \ldots$

Step 1—Parenthesis	...
Step 2—Exponents	$5 + 4 \bullet 3/2 - 1$
Step 3—Multiply and Divide	$5 + 12/2 - 1$
	$5 + 6 - 1$
Step 4—Add and Subtract	$11 - 1$
	10

Pop Quiz: *Four Fours*   (page 93)

$$4 \;/\; 4 \;+\; 4 \;-\; 4 \;=\; 1$$
$$4 \;/\; 4 \;+\; 4 \;/\; 4 \;=\; 2$$
$$(4 \;+\; 4 \;+\; 4) \;/\; 4 \;=\; 3$$
$$4 \;+\; (4 \;-\; 4) \;\bullet\; 4 \;=\; 4$$
$$(4 \;\bullet\; 4 \;+\; 4) \;/\; 4 \;=\; 5$$
$$(4 \;+\; 4) \;/\; 4 \;+\; 4 \;=\; 6$$
$$4 \;+\; 4 \;-\; 4 \;/\; 4 \;=\; 7$$
$$4 \;+\; 4 \;+\; 4 \;-\; 4 \;=\; 8$$
$$4 \;+\; 4 \;+\; 4 \;/\; 4 \;=\; 9$$

Pop Quiz: *Expressions vs. Equations*   (page 99)

How many valid equations can be created from these four expressions? Well, each of the arrows below connect two expressions that together make a valid equation. You'll notice that there isn't an arrow connecting 14 and 7. Why? Because $14 = 7$ is not a valid equation! Which means that we can make a total of five valid equations from our four expressions.

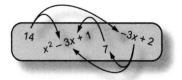

And here are those five equations:

$x^2 - 3x + 1 = 14$	$-3x + 2 = 14$
$x^2 - 3x + 1 = -3x + 2$	$-3x + 2 = 7$
$x^2 - 3x + 1 = 7$	

But wait! Aren't there five more equations? Can't we switch the expressions on the left and right side of these equations so that, for example, we turn the first equation into $14 = x^2 - 3x + 1$? Well, that equation is perfectly valid, but it's not unique. In fact, it means the exact same thing as $x^2 - 3x + 1 = 14$ . . . so there's no reason to include it.

Algebra Tutorial: *How to Solve an Equation* (page 107)

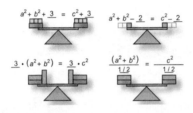

Pop Quiz: *Let's Solve Some Equations!* (page 116)

> **The 4 Steps of Equation Solving**
> 1. Simplify both sides of the equation.
> 2. Move all parts of the equation that contain the variable you're solving for to the same side.
> 3. Isolate the variable using multiplication, division, exponentiation, or by taking roots.
> 4. Check your solution!

1. $2 + x - 2 \cdot 5 = 4/2 - x$

    Step 1:

        Simplify $2 \cdot 5$ and $4/2$: $2 + x - 10 = 2 - x$

        Simplify $2 - 10$: $-8 + x = 2 - x$

    Step 2:

        Add x to both sides: $-8 + x + x = 2 - x + x$

        Simplify: $-8 + 2x = 2$

Add 8 to both sides: $-8 + 2x + 8 = 2 + 8$

Simplify: $2x = 10$

Step 3:

Divide both sides by 2: $2x/2 = 10/2$

Simplify: $x = 5$

Step 4:

Plug-in and check LHS: $2 + 5 - 2 \bullet 5 = 2 + 5 - 10 = 7 - 10 = -3$

Plug-in and check RHS: $4/2 - 5 = 2 - 5 = -3$

LHS $\rightarrow$ -3 and RHS $\rightarrow$ -3 ✓

2. $x/3 + 3 = x/2$

The fact that $3 \bullet (a + b) = 3a + 3b$ (which we'll cover in detail in chapter 4) will come in handy for this problem.

Step 1:

Multiply both sides by 3: $(3x/3) + 3 \bullet 3 = 3x/2$

Simplify $3x/3$ and $3 \bullet 3$: $x + 9 = 3x/2$

Multiply both sides by 2: $2x + 18 = 3x$

Step 2:

Subtract $2x$ from both sides: $2x + 18 - 2x = 3x - 2x$

Simplify: $18 = x$

Step 3: Nothing to do!

Step 4:

Plug-in and check LHS: $18/3 + 3 = 6 + 3 = 9$

Plug-in and check RHS: $18/2 = 9$

LHS $\rightarrow$ 9 and RHS $\rightarrow$ 9 ✓

3. $\sqrt{x} - 4 = -\sqrt{x}$

Step 1: Nothing to do!

Step 2:

Add $\sqrt{x}$ to both sides: $\sqrt{x} - 4 + \sqrt{x} = -\sqrt{x} + \sqrt{x}$

Simplify: $2\sqrt{x} - 4 = 0$

Add 4 to both sides: $2\sqrt{x} - 4 + 4 = 0 + 4$

Simplify: $2\sqrt{x} = 4$

Step 3:

Divide both sides by 2: $2\sqrt{x}/2 = 4/2$

Simplify: $\sqrt{x} = 2$

Square both sides: $(\sqrt{x})^2 = 2^2$

Simplify: $x = 4$

Step 4:

    Plug-in and check LHS: $\sqrt{4} - 4 = 2 - 4 = -2$

    Plug-in and check RHS: $-\sqrt{4} = -2$

    LHS $\rightarrow -2$ and RHS $\rightarrow -2$ ✓

4. $2x^2 - 3 + x^2 = 9 + x^2$

    Step 1:

        Combine $2x^2$ and x^2: $3x^2 - 3 = 9 + x^2$

    Step 2:

        Subtract x^2 from both sides: $3x^2 - 3 - x^2 = 9 + x^2 - x^2$

        Simplify: $2x^2 - 3 = 9$

        Add 3 to both sides: $2x^2 - 3 + 3 = 9 + 3$

        Simplify: $2x^2 = 12$

    Step 3:

        Divide both sides by 2: $2x^2/2 = 12/2$

        Simplify: $x^2 = 6$

        Take the square root of both sides: $\sqrt{x^2} = \pm\sqrt{6}$ (remember both solutions!)

        Simplify: $x = \pm\sqrt{6}$

    Step 4:

        Check solution $x = \sqrt{6}$

            Plug-in and check LHS: $2(\sqrt{6})^2 - 3 + (\sqrt{6})^2 = 2 \cdot 6 - 3 + 6 = 15$

            Plug-in and check RHS: $9 + (-\sqrt{6})^2 = 9 + 6 = 15$

            LHS $\rightarrow 15$ and RHS $\rightarrow 15$ ✓

        Check solution $x = -\sqrt{6}$

            Plug-in and check LHS: $2(-\sqrt{6})^2 - 3 + (-\sqrt{6})^2 = 2 \cdot 6 - 3 + 6 = 15$

            Plug-in and check RHS: $9 + (-\sqrt{6})^2 = 9 + 6 = 15$

            LHS $\rightarrow 15$ and RHS $\rightarrow 15$ ✓

Secret Agent Math-Lib: *Where Are My Keys?*   (page 117)

Here's your decoded secret message:

<u>L</u>   <u>O</u>   <u>O</u>   <u>K</u>

<u>I</u>   <u>N</u>

<u>Y</u>   <u>O</u>   <u>U</u>   <u>R</u>

<u>S</u>   <u>H</u>   <u>O</u>   <u>E</u> !

If you need help solving any of the problems, you can use the letters in this message to work backwards and find the solution to the problem you're stuck on.

Final Exam   (page 130)

- ## The Building Blocks of Algebra: Variables, Expressions, and Equations

1. Equation: $2x - 3 = 7$ with $x \to 5$
2. Equation: $15 - x^2 = -2x$ with $x \to -3$
3. Equation: $3\sqrt{x} - 6 = x - 6$ with $x \to 9$

   Be sure to check each of these answers by plugging the solutions for x into both the left and right sides of the equations and ensuring that LHS = RHS!

- ## Order of Operations

4. $4 \cdot (4 / 4 + 4) = 20$
5. $4 \cdot 4 + 4 + 4 = 24$
6. $(4 + 4) \cdot 4 - 4 = 28$
7. $(4 \cdot 4) + (4 \cdot 4) = 32$
8. $(4 + 4) \cdot 4 + 4 = 36$

- ## Solving Equations

The three equations that we **can** solve and their solutions are:

9. $-3x + 2 = 14$

Subtract 2 from both sides:	$-3x + 2 - 2 = 14 - 2$
Simplify:	$-3x = 12$
Divide both sides by -3:	$-3x / -3 = 12 / -3$
Simplify:	$x = -4$
Plug in and check:	$-3 \cdot -4 + 2 = 12 + 2 = 14$ ✓

10. $-3x + 2 = 7$

Subtract 2 from both sides:	$-3x + 2 - 2 = 7 - 2$
Simplify:	$-3x = 5$
Divide both sides by -3:	$-3x / -3 = 5 / -3$

Simplify: $x = -5/3$

Plug in and check: $-3 \cdot -5/3 + 2 = 5 + 2 = 7$ ✓

11. $x^2 - 3x + 1 = -3x + 2$

Add $3x$ to both sides: $x^2 - 3x + 1 + 3x = -3x + 2 + 3x$

Simplify: $x^2 + 1 = 2$

Subtract 1 from both sides: $x^2 + 1 - 1 = 2 - 1$

Simplify: $x^2 = 1$

Square root of both sides: $\sqrt{x^2} = \pm\sqrt{1}$

Simplify: $x = \pm 1$ (remember both solutions!)

Check solution $x = 1$:

   Plug in and check LHS: $1^2 - 3 \cdot 1 + 1 = 1 - 3 + 1 = -1$

   Plug in and check RHS: $-3 \cdot 1 + 2 = -3 + 2 = -1$

   LHS $\rightarrow -1$ and RHS $\rightarrow -1$ ✓

Check solution $x = -1$:

   Plug in and check LHS: $(-1)^2 - 3 \cdot (-1) + 1 = 1 + 3 + 1 = 5$

   Plug in and check RHS: $-3 \cdot (-1) + 2 = 3 + 2 = 5$

   LHS $\rightarrow 5$ and RHS $\rightarrow 5$ ✓

And the two equations that we *cannot* solve are:

$$x^2 - 3x + 1 = 14$$
$$x^2 - 3x + 1 = 7$$

Why not? Well, try it for yourself if you haven't. You'll find that none of the techniques you've seen in this chapter will help. We need to learn more. Which means that there's still a lot for us to learn about algebra. Interestingly, this type of equation (called a quadratic equation) is exactly the type of equation that we need to be able to solve to answer the "challenge problems" from the end of the first chapter. Which means that we'll definitely end up knowing how to answer these . . . you'll just have to wait a bit longer.

• Solving Algebra Problems

12. The equation we need to plot is shown on the right side (with the numerical values of the constants W_{towels} and W_{toy} already plugged in). Below the equation, you can see a table showing the values of W_{total} that we get for each value of N_{toys}. Each row of this table represents an ordered (N_{toys}, W_{total}) pair that you can see plotted on the left. All the ordered pairs fall along a line that extends from (0, 2) to (10, 7). As we'll find out in the next

chapter, this is an example of what's called a "linear equation" since all of the points fall along a straight line.

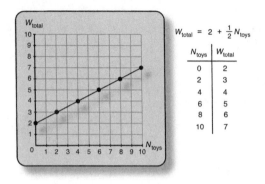

$$W_{total} = 2 + \frac{1}{2} N_{toys}$$

N_{toys}	W_{total}
0	2
2	3
4	4
6	5
8	6
10	7

CHAPTER 3: WALK THE NUMBER LINE

Pop Quiz: *Ancient Arithmetic*   (page 139)

These numbers are just examples. Your numbers may be very different . . . but the general idea will be the same!

1. Think of a number greater than 1,000 but less than 10,000......___658___
   Now translate this into Egyptian hieroglyphics:

2. Think of a smaller number that's still greater than 1,000...........___537___
   Now translate that into Egyptian hieroglyphics:

3. Add the two numbers together . . . Egyptian-style:

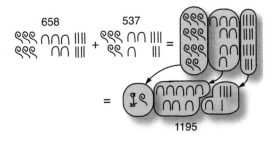

4. Now subtract the second number from the first . . . Egyptian-style:

Hieroglyphic addition and subtraction works, but it's kind of a pain. Multiplication and division are even worse. All in all, I'm thankful for our modern decimal system, but it's interesting to think about other perfectly effective methods for doing the math that you know and love . . . including algebra.

Pop Quiz: *Algebra, Decimal Numbers, and Addition*   (page 142)

These numbers are just an example. Feel free to substitute your own!

I'll think of a number between 100 and 1,000 .... __451__ (#1)
Now you think of another number between 100 and 1,000 .... __877__ (#2)

Let's write each of these as the sum of three numbers times different multiples of 10:

#1: __451__ = __4__ • 100 + __5__ • 10 + __1__
#2: __877__ = __8__ • 100 + __7__ • 10 + __7__

Now add up both numbers in each of the three columns for the hundreds, tens, and ones place. Write them below in the corresponding column.

#1 + #2 = __12__ • 100 + __12__ • 10 + __8__ = __1200 + 120 + 8 = 1328__

What do you think? Is adding numbers by powers of ten like this potentially faster than using the traditional right-to-left "carrying" method? Of course, once you're good at it, you don't have to write it all out like this . . . you can just do it in your head.

Pop Quiz: *Absolute Values on the Number Line*   (page 147)

In this solution I have chosen $x=0$ and $y=-5$, but you could have chosen any values.

1. $|-4| =$ __**4**__
2. $|3| =$ __**3**__
3. $|-4+3| =$ __**1**__
4. $|-4|+|3| =$ __**7**__
5. $|-4-3| =$ __**7**__
6. $|x+3| = |0+3| = |3| =$ __**3**__
7. $|x-y| = |0-(-5)| = |5| =$ __**5**__

The absolute values of -4 and 3 tell you how far each number is from zero. You can think of the absolute value $|-4+3|$ as being equal to the net distance you travel away from the origin if you start at 0, walk 4 steps in the negative direction, and then 3 steps in the positive direction. You can think of the absolute value $|-4-3|$ as the distance between the points -4 and 3. The meaning of $|x+3|$ is similar to that of $|-4+3|$, and the meaning of $|x-y|$ is similar to that of $|-4-3|$. We talk more about this in the following sections.

Secret Agent Math-Lib: *What's the Secret to a Safe Commute?*   (page 150)

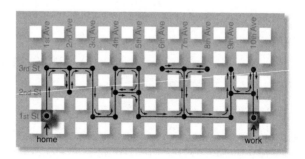

Pop Quiz: *Writing Distances on the Number Line*   (page 154)

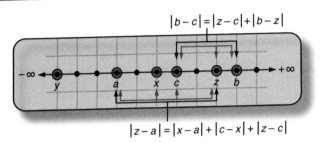

1. $|b-c| = |z-c| + |b-z|$
2. $|z-a| = |x-a| + |c-x| + |z-c|$

Pop Quiz: *From Fraction to Decimal*   (page 159)

1. $^{19}\!/_{20} =$ ___0.95___
2. $^{7}\!/_{9} =$ ___$0.\overline{7}$___
3. $^{72}\!/_{195} =$ ___$0.\overline{3692307}$___

Pop Quiz: *Name that Number*   (page 163)

1. $^{10}\!/_{3}$ is a **rational** number since it's a fraction written as the ratio of two integers.
2. $\sqrt{3}$ is an **irrational** number which means that it's decimal digits never stop or repeat.
3. -17 is an **integer**.
4. $\sqrt{64}$ is an **integer** once it is simplified to 8.
5. π is a (very famous) **irrational** number.

Pop Quiz: *Linear or Not?*   (page 164)

1. $3 - x = x \rightarrow$ This equation is **linear** since x isn't raised to a power other than 1.
2. $5x^2 + 8 = 0 \rightarrow$ This equation is **not linear** since the variable x is squared.
3. $x \bullet x - 9 = 0 \rightarrow$ Use $x \bullet x = x^2$ to simplify to $x^2 - 9 = 0$, which is **not linear**.

Pop Quiz: *English to Equation Translation*   (page 166)

Here's the sentence we need to translate:

> *The mean of 10 and some number larger than 10 is equal to 1 more than the range of these two numbers.*

And here's how to actually translate it into an equation:

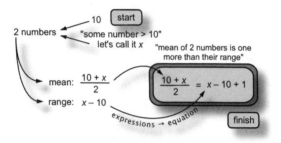

Our final equation is over there on the right. The mean of our two numbers, *x* and 10, is on the left side of the equals sign, and the range of the two numbers is on the right. You may have wondered why I've written the range as *x* – 10 instead of 10 – *x*. Well, the range tells us how big of a gap exists between two numbers, which means that it's always a positive number. Since the problem states that the number we've called *x* is greater than 10, writing the range as *x* – 10 ensures that the range will be positive.

Pop Quiz: *Let's Solve a Single Variable Linear Equation!*   (page 173)

The English to equation translation of

> *The mean of 10 and some number less than 10 is equal to 1 more than the range of these two numbers.*

is the equation

$$\frac{10+x}{2} = 10 - x + 1$$

Notice that the range in this equation is $10 - x$ (instead of $x - 10$ as it was before) since we are told that $x < 10$. We can now use our five-step process

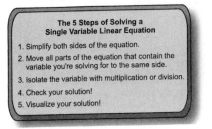

**The 5 Steps of Solving a
Single Variable Linear Equation**

1. Simplify both sides of the equation.
2. Move all parts of the equation that contain the variable you're solving for to the same side.
3. Isolate the variable with multiplication or division.
4. Check your solution!
5. Visualize your solution!

to solve this equation. First, we can simplify the pieces on the right side of the equation to find that $10 - x + 1 = 11 - x$. Then, we can multiply both sides of the equation by 2 to get

$$10 + x = 2 \bullet 11 - 2x$$

which simplifies to

$$10 + x = 22 - 2x$$

Next we need to move the parts of the equation containing x to the left side, and the parts with constants to the right . . . meaning we need to add $2x$ and subtract 10 from both sides:

$$10 + x + 2x - 10 = 22 - 2x + 2x - 10$$

After simplifying, we get

$$3x = 12$$

If we divide both sides of the equation by 3, we get the solution $x = 4$. Before finishing, you should try plugging this solution back into the original equation to make sure it works. Here is the number line showing the solution to this problem:

And here is how you can find this solution graphically!

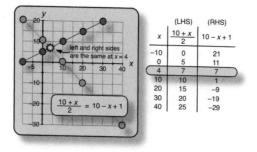

x	(LHS) $\dfrac{10+x}{2}$	(RHS) $10-x+1$
−10	0	21
0	5	11
4	7	7
10	10	1
20	15	−9
30	20	−19
40	25	−29

Pop Quiz: *Let's Solve an Absolute Value Equation!*   (page 179)

Let's solve the absolute value equation $|x-3|-1=2$ using our four-step process

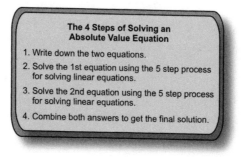

The 4 Steps of Solving an Absolute Value Equation

1. Write down the two equations.
2. Solve the 1st equation using the 5 step process for solving linear equations.
3. Solve the 2nd equation using the 5 step process for solving linear equations.
4. Combine both answers to get the final solution.

First, let's break the absolute value equation up into the two equations we get when the absolute value is either positive or negative.

1. When $x \geq 3$, the absolute value is positive, and therefore the equation we need to solve is $x-3-1=2$. We can solve this equation by simply moving all of the constants to the right side of the equals sign. The solution is **$x=6$**.

2. When $x<3$, the absolute value is negative, and therefore the equation we need to solve is $-(x-3)-1=2$. Let's first add 1 to both sides to get $-(x-3)=3$. Now let's multiply both sides by −1 to get $x-3=-3$. Finally, let's add 3 to both sides to get the final solution: **$x=0$**.

So the equation $|x-3|-1=2$ has two solutions: $x=6$ and $x=0$. You should try plugging both of these solutions into the original equation to make sure they work. Here is the number line showing the solutions to this problem:

Pop Quiz: *Equality vs. Inequality* (page 180)

1. What greater-than-20 numbers can you pair up with 20 so that the mean of the pair is always greater than 3 less than their range?

$$\frac{20+x}{2} > x - 20 - 3$$

2. What greater-than-5 numbers can you pair up with 5 so that the mean of the pair is always less than or equal to 2 more than their range?

$$\frac{5+x}{2} \leq x - 5 + 2$$

3. What greater-than-10 numbers can you pair up with 10 so that the mean of the pair is always greater than or equal to 1 more than their range?

$$\frac{10+x}{2} \geq x - 10 + 1$$

Pop Quiz: *Inequality to Interval Notation Translation* (page 187)

1. $-1 < x \leq 1$ .... **(−1, 1]**

2. $4 \leq x < 14$ .... **[4, 14)**

3. $10 < x$ .... **(10, + ∞)**

4. $x \leq -9$ .... **(−∞, −9]**

Final Exam (page 194)

• Numbers, the Number Line, and Absolute Values

1. This is one of many possible solutions to this problem. The equation says that the distance between x and y can be found by subtracting the distance between c and x from the distance between c and y.

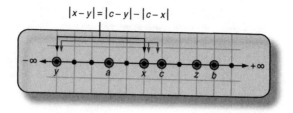

$$|x - y| = |c - y| - |c - x|$$

• Solving Linear Equations

2. The English to equation translation of:

> The mean of 10 and some number greater than 10 is 1 less than the range of these two numbers.

gives us the equation

$$\frac{10 + x}{2} = x - 10 - 1$$

We can solve this equation by combining the pieces on the right side to get $x - 10 - 1 = x - 11$, and then multiplying both sides by 2:

$$10 + x = 2x - 2 \bullet 11$$

This simplifies to give us

$$10 + x = 2x - 22$$

Next we need to move the parts of the equation containing x to the right side, and the parts with constants to the left . . . meaning we need to subtract x and add 22 from both sides:

$$10 + x - x + 22 = 2x - 22 - x + 22$$

After simplifying, we get

$$32 = x$$

So our solution is $x = 32$. Before finishing, you should try plugging this solution back into the original equation to make sure it works. Finally, here is the number line showing the solution to the problem:

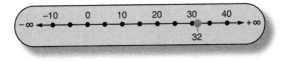

• Solving Absolute Value Equations

3.

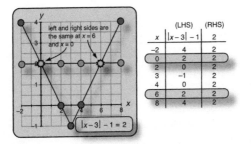

• Solving Inequalities

4. Let's start solving the inequality

$$\frac{10+x}{2} \geq |x-10|+1$$

by writing down the two inequalities given by the two possible values of the absolute value. First, when $x-10 \geq 0$, we get the inequality

$$\frac{10+x}{2} \geq x-10+1$$

We can solve this inequality by simplifying the right side and multiplying both sides by 2. The result is

$$10+x \geq 2x-18$$

Now, if we subtract x and add 18 to both sides, we get

$28 \geq x$

We can turn this around to write $x \leq 28$. When $x - 10 \leq 0$ in the original inequality, we get our second inequality to solve:

$$\frac{10 + x}{2} \geq -(x - 10) + 1$$

If we again simplify the right side and multiply both sides by 2, we get

$10 + x \geq -2x + 22$

Adding $2x$ and subtracting 10 from both sides gives

$$3x \geq 12$$

And dividing both sides by 3 gives us

$x \geq 4$

This means that the inequality

$$\frac{10 + x}{2} \geq |x - 10| + 1$$

is satisfied for any value of x that is both greater than or equal to 4 and less than or equal to 28. You should try plugging in some numbers to make sure this is true! This solution can be written in interval notation like this:

[4, 28]

Finally, here's the complete solution on the number line:

CHAPTER 4: ARITHMETIC 2.0

Pop Quiz: *Let's Add and Subtract Some Fractions!*   (page 203)

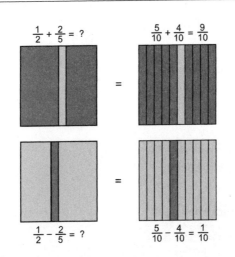

$$\frac{1}{2} + \frac{2}{5} = ?$$ $$\frac{5}{10} + \frac{4}{10} = \frac{9}{10}$$

$$\frac{1}{2} - \frac{2}{5} = ?$$ $$\frac{5}{10} - \frac{4}{10} = \frac{1}{10}$$

Pop Quiz: *Let's Multiply "n" Fractions!*   (page 206)

We can write a rule that tells us how to multiply *n* fractions together by writing each fraction with a subscript on both its numerator and denominator from 1, 2, 3, and on up to *n* . . . which can be any number we want:

$$\frac{a_1}{b_1} \cdot \frac{a_2}{b_2} \cdot \frac{a_3}{b_3} \cdot \ldots \cdot \frac{a_n}{b_n} = \frac{a_1 a_2 a_3 \cdots a_n}{b_1 b_2 b_3 \cdots b_n}$$

Pop Quiz: *Fractions, Arithmetic, and Algebra*   (page 207)

1. $\frac{1}{2} + \frac{1}{3} = x$

       Multiply both sides by 2→ $1 + \frac{2}{3} = 2x$

       Multiply both sides by 3→ $3 + 2 = 6x$

       Simplify left side→ $5 = 6x$

       Divide both sides by 6→ $x = \frac{5}{6}$

2. $\frac{1}{2} - \frac{1}{3} = x$

      Multiply both sides by 2 → $1 - \frac{2}{3} = 2x$

      Multiply both sides by 3 → $3 - 2 = 6x$

      Simplify left side → $1 = 6x$

      Divide both sides by 6 → $x = \frac{1}{6}$

3. $\frac{1}{2} \cdot \frac{1}{3} = x$

      Multiply both sides by 2 → $1 \cdot \frac{1}{3} = 2x$

      Multiply both sides by 3 → $1 \bullet \; 1 = 6x$

      Simplify left side → $1 = 6x$

      Divide both sides by 6 → $x = \frac{1}{6}$

4. $\frac{1}{2} \; / \; \frac{1}{3} = x$

      Multiply both sides by $\frac{1}{3}$ → $\frac{1}{2} = \frac{1}{3} \cdot x$

      Multiply both sides by 3 → $x = \frac{3}{2}$

Pop Quiz: *What's Your Sign?*   (page 210)

1. $2 \bullet -3 \bullet 3 / -5 / -1$ .... **negative**

2. $1 \bullet -1 / 1 \bullet 1 \bullet -1$ .... **positive**

3. $-7 \bullet -4 / -3 \bullet -8$ .... **positive**

Pop Quiz: *Positive and Negative Outcomes . . . Part Two*   (page 213)

$|x| < |y|$	$x > 0$ $y > 0$	$x > 0$ $y < 0$	$x < 0$ $y > 0$	$x < 0$ $y < 0$
$x + y$	positive	negative	positive	negative
$x - y$	negative	positive	negative	positive
$x \cdot y$	positive	negative	negative	positive
x / y	positive	negative	negative	positive

Pop Quiz: *What Do You Get?*   (page 217)

The only thing that changes when we move from $y > 0$ to $y < 0$ are the signs of $x \cdot y$ (they're all the exact opposite). The magnitude of $x \cdot y$ does not change.

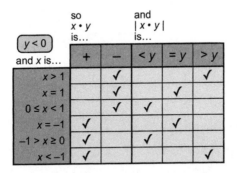

	so $x \cdot y$ is…		and $|x \cdot y|$ is…		
$y < 0$ and x is…	$+$	$-$	$< y$	$= y$	$> y$
$x > 1$		✓			✓
$x = 1$		✓		✓	
$0 \le x < 1$		✓	✓		
$x = -1$	✓			✓	
$-1 > x \ge 0$	✓		✓		
$x < -1$	✓				✓

Pop Quiz: *Let's Simplify Some Expressions!*   (page 223)

1. $2x \cdot 3x - (10z / 5z)$

   The first thing to do is gather constants and variables in each term, like this:

   $(2 \cdot 3)(x \cdot x) - (10 / 5)(z / z)$

   Then we just need to simplify to get

   $6x^2 - 2$

2. $8z / 2x + 4y$

   Again, just gather constants and variables in each term:

   $(8 / 2)(z / x) + 4y$

   Notice that there's nothing to gather in the second term—it's already simplified. But we still need to simplify the first term:

   $4z / x + 4y$

3. $7z \cdot 4x / 7y + 3xy / 7x - 9x$

Gathering constants and variables in this three term problem gives us

$(7 \cdot 4 / 7)(z \cdot x / y) + (3 / 7)(xy / x) - 9x$

Simplifying, we get

$4zx / y + 3y / 7 - 9x$

If you're not sure why we're able to swap numbers around like this to change up the order in which we multiply and divide, check out the Math Properties section.

Pop Quiz: *How to Add Quickly* (page 224)

3, 9, 2, 6, 4, 2, 7, 5, 1, 9, 3

Remember, the trick is to look for pairs or groups of numbers that add to 10. There are lots of groupings you could come up with. Here's one possibility:

$(3 + 7) + (9 + 1) + (6 + 4) + (2 + 3 + 5) + 9 + 2 = 4 \cdot 10 + 11 = 51$

Pop Quiz: *Which Property Lets You Do This?* (page 228)

1. $1 + (2 + 3 + 4) = (1 + 2 + 3) + 4$ .... **AA**
2. $15 + 2 + 7 = 15 + 7 + 2$ .... **CA**
3. $2 \cdot (4 - 5) - 1 = 2 \cdot 4 - 2 \cdot 5 - 1$ .... **D**
4. $(5 \cdot 3) \cdot 1 + 7 = 5 \cdot (3 \cdot 1) + 7$ .... **AM**
5. $8 + 3 \cdot 8 = 8 \cdot 3 + 8$ .... **CA/CM**

The last problem requires both the commutative properties of addition and multiplication since the two terms are switched and the order of the numbers being multiplied is also switched.

Pop Quiz: *How to Multiply Quickly*   (page 231)

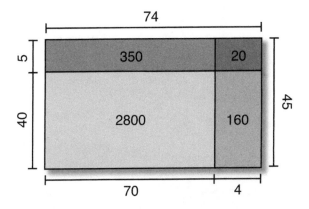

If we split 74 into $70 + 4$ and 45 into $40 + 5$, the problem becomes

$74 \bullet 45 = (70 + 4) \bullet (40 + 5)$

Using FOIL, we get

$74 \bullet 45 = (70 \bullet 40) + (70 \bullet 5) + (4 \bullet 40) + (4 \bullet 5)$

These are all fairly easy to solve in your head, which means you can quickly calculate

$74 \bullet 45 = 2800 + 350 + 160 + 20 = 3330$

Pop Quiz: *Let's Combine Some Like Terms!*   (page 239)

1. $10x + 5xy + 3z + 2 \bullet (xy + z)$

   First we need to distribute the 2 in the final term

   $10x + 5xy + 3z + 2xy + 2z$

   No further simplification is needed, so let's now group the like terms

   $10x + (5xy + 2xy) + (3z + 2z)$

Finally, let's combine the like terms to get our final simplified expression

$10x + 7xy + 5z$

2. $x^2 - xz^2 \bullet (x/z^2 - y) - 2z^2 \bullet xy$

First we need to distribute the second term

$x^2 - xz^2 \bullet x/z^2 - xz^2 \bullet (-y) - 2z^2 \bullet xy$

Now we need to simplify the terms

$x^2 - x^2 + xz^2y - 2xz^2y$

Let's now group the like terms

$(x^2 - x^2) + (xz^2y - 2xz^2y)$

And then combine them to get our final simplified expression

$-xz^2y$

3. $2x - (x + y^2 + 3xy)$

First we need to distribute the negative sign through the terms in parentheses. If this seems strange to you, think of the expression like this instead

$2x + (-1) \bullet (x + y^2 + 3xy)$

Then it's clear that we need to distribute like this

$2x - x - y^2 - 3xy$

No further simplification is needed, so let's group the like terms

$(2x - x) - y^2 - 3xy$

And now let's combine them to get our final simplified expression

$x - y^2 - 3xy$

Pop Quiz: *Even and Odd Exponentiation*   (page 242)

Any even number raised to either an even or odd power always produces an even number. For example, $2^2 = 4$ and $2^3 = 8$ are both even. On the other hand, any odd number raised to either an even or odd power always produces an odd number. For example, $3^2 = 9$ and $3^3 = 27$ are both odd.

Pop Quiz: *Let's Multiply Some Expressions . . .*
with Exponents!   (page 244)

1. $5^2 \bullet 3^8 \bullet 5$ $\rightarrow$ $(5 \bullet 5^2) \bullet 3^8 = (5^1 \bullet 5^2) \bullet 3^8 = 5^{1+2} \bullet 3^8 = 5^3 \bullet 3^8$

2. $3^2 \bullet x^2 \bullet x^8$ $\rightarrow$ $3^2 \bullet (x^2 \bullet x^8) = 3^2 \bullet x^{2+8} = 3^2 \bullet x^{10}$

3. $4^2 \bullet 3^5$ $\rightarrow$ $4^2 \bullet 3^5$ (can't simplify any further since the bases are different)

4. $x \bullet (x^2 + x^3)$ $\rightarrow$ $x \bullet x^2 + x \bullet x^3 = x^1 \bullet x^2 + x^1 \bullet x^3 = x^{1+2} + x^{1+3} = x^3 + x^4$

Pop Quiz: *Let's Divide Some Expressions . . . with*
Exponents!   (page 246)

1. $2^2 \bullet 3^8 / 3^5$ $\rightarrow$ $2^2 \bullet (3^8 / 3^5) = 2^2 \bullet 3^{8-5} = 2^2 \bullet 3^3$

2. x^6 / x^3 $\rightarrow$ $x^{6-3} = x^3$

3. $4^2 / 3^5$ $\rightarrow$ $4^2 / 3^5$ (can't simplify any further since the bases are different)

4. $x \bullet (x^2 / x^3)$ $\rightarrow$ $x \bullet (x^{2-3}) = x \bullet x^{-1} = x^{1-1} = x^0 = 1$

Pop Quiz: *Arithmetic with Exponents*   (page 248)

1. $(xy)^2 \bullet x^2 / x^3$ $\rightarrow$ $x^2 y^2 \bullet x^2 / x^3 = x^2 y^2 \bullet x^{2-3} = x^2 y^2 \bullet x^{-1} = y^2 \bullet x^{2-1} = xy^2$

2. $x \bullet (x^3 / x^2)^2$ $\rightarrow$ $x \bullet (x^{3-2})^2 = x \bullet (x^1)^2 = x \bullet x^{1 \bullet 2} = x \bullet x^2 = x^{1+2} = x^3$

3. $x \bullet [\, xy - (xy)^2 \,]$ $\rightarrow$ $x \bullet [\, xy - x^2 y^2 \,] = x \bullet xy - x \bullet x^2 y^2 = x^2 y - x^3 y^2$

Pop Quiz: *Find the Exponent*   (page 251)

We need to figure out what n must be in this equation

$[\, (x^2 \bullet 2^3 x)^2 \,]^n = 2x$

Let's start by simplifying the inside part of the left side:

$$(x^2 \bullet 2^3 x)^2 = (2^3 x^3)^2 = 2^6 x^6$$

Now, let's put the equation back together

$$[\, 2^6 x^6 \,]^n = 2x$$

In order for the left side to equal the right, you can now see that $n = \dfrac{1}{6}$. If you plug that value in, the left side becomes

$$[\, 2^6 x^6 \,]^{1/6} = 2^{6 \bullet 1/6} x^{6 \bullet 1/6} = 2x$$

Which means that $2x$ is the sixth root of $(x^2 \bullet 2^3 x)^2$.

Pop Quiz: *Arithmetic with Roots*   (page 254)

1. $[\, (x^2 y)^2 \bullet x / x^3 \,]^{1/2}$ $\rightarrow$ $[\, x^4 y^2 \bullet x^{1-3} \,]^{1/2} = [\, y^2 x^4 \bullet x^{-2} \,]^{1/2} = [\, y^2 x^2 \,]^{1/2} = xy$
2. $[\, x \bullet (x^3 / x^2)^2 \,]^{1/3}$ $\rightarrow$ $[\, x \bullet (x^{3-2})^2 \,]^{1/3} = [\, x \bullet x^2 \,]^{1/3} = [\, x^3 \,]^{1/3} = x$
3. $[\, x \bullet (\, 1 - x^2 \,) \,]^{1/4}$ $\rightarrow$ $[\, x - x^3 \,]^{1/4}$

Notice that we've simplified the third problem as much as we can. Remember, the fourth property of roots says that $[\, x - x^3 \,]^{1/4} \neq x^{1/4} - x^{3/4}$.

Pop Quiz: *Combining Roots and Powers*   (page 256)

We can calculate the value of the expression $9^{3/2}$ using any of these three methods:

1. Power before the root: $9^{3/2} = (9^3)^{1/2} = 729^{1/2} = 27$
2. Root before the power: $9^{3/2} = (9^{1/2})^3 = 3^3 = 27$
3. Divide the exponent $\dfrac{3}{2}$ and then use a calculator to find: $9^{1.5} = 27$

Secret Agent Math-Lib: *Who Carved the Mystery Message?*   (page 257)

1. Which expression is equal to x^2? .... <u>**5**</u> • 5 = <u>**25**</u>

2. Which expression will be odd for any value of x? .... <u>**4**</u> + 11 = <u>**15**</u>

3. Which expression is equal to x_{-1}? .... <u>**1**</u> + 20 = <u>**21**</u>

4. Which expression requires that $x > 0$? .... <u>**2**</u> • 2 = <u>**4**</u>

5. Which expression is equal to $\sqrt[3]{x^2}$? .... <u>**3**</u> • 3 = <u>**9**</u>

Who carved that mysterious message?

$$\frac{Y}{(1)} \quad \frac{O}{(2)} \quad \frac{U}{(3)} \qquad \frac{D}{(4)} \quad \frac{I}{(5)} \quad \frac{D}{(4)}$$

Final Exam   (page 260)

• Arithmetic 1.0: Integers and Fractions

1.

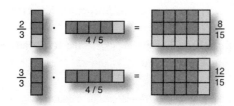

Notice that the product of the numerators is equal to the number of dark squares, and the product of the denominators (a common denominator) is equal to the total number of squares. You can also see by looking at the figures on the right whether or not the fraction has an equivalent "reduced" version. For example, you can tell that the fraction $^{12}/_{15}$ can be reduced to $^{4}/_{5}$ since each of the three rows is exactly the same!

• Arithmetic 2.0: Variables and Math Properties

There are many possible examples for this. Here are a few . . .

2. Let's think about a couple of things and then figure out which of them are commutative. How about pouring a cup of coffee with milk? Or putting on your socks and shoes? Most people would agree that it doesn't matter whether you first pour the coffee and then the milk, or vice versa—the end result is a cup of coffee with milk. But the order in which you

put on your socks and shoes makes a big difference. Socks before shoes gets you ready for a happy day of walking, but shoes before socks just gets you a couple of destroyed socks and a bunch of quizzical looks.

3. Imagine you're mixing concrete made from cement mix, gravel, and water. The instructions say to first mix the gravel with the cement mix, and then to add the water. But you accidentally mixed the water and cement mix first, waited a while, and then tried to mix in the gravel . . . after the cement had set. Mixing cement is *not* associative!

4. One real-world example of the distributive property in action happens when you double a recipe. Let's say you're making a cake that needs two ingredients . . . which we'll call *a* and *b*. A finished batch of this recipe ready for baking therefore looks like $a + b$. Now, if you want to double the recipe you have two options. First, you could combine twice as much of each ingredient all at once to get a batter that looks like $2a + 2b$. Or, you could make two separate batches of your $a + b$ recipe, and then combine them together to get $2 \cdot (a + b)$. Either way, the cake is going to taste the exact same . . . which means that $2 \cdot (a + b) = 2a + 2b$.

• Exponentiation

5. Here's how to multiply and divide the numbers 2,000 and 2,000,000 using scientific notation:

• First write the number 2,000 in scientific notation (you can use the example of 3,000 above to help) . . . . **2×10^3**

• Next write the number 2,000,000 in scientific notation . . . .
    2×10^6

• Now multiply these together . . . .

   $2 \times 10^3 \cdot 2 \times 10^6 = (2 \cdot 2) \cdot 10^3 \cdot 10^6 = 4 \cdot 10^{3+6} = 4 \times 10^9$ (which is 4 billion)

• And finally divide them . . . .

   $2 \times 10^3 / 2 \times 10^6 = (2 / 2) \cdot 10^3 / 10^6 = 1 \cdot 10^{3-6} = 1 \times 10^{-3}$ (which is 1/1000)

• Taking Roots . . . Square, Cube, and Otherwise

6. Let's use a calculator to check the accuracy of these approximations to $2^{\sqrt{2}}$:

• Using two products gives us: $2^{\sqrt{2}} \approx 2 \cdot 2^{0.4} \approx 2.639\ldots$

- Using three products gives us: $2^{\sqrt{2}} \approx 2 \cdot 2^{0.4} \cdot 2^{0.01} \approx 2.657...$

- Using four products gives us: $2^{\sqrt{2}} \approx 2 \cdot 2^{0.4} \cdot 2^{0.01} \cdot 2^{0.004} \approx 2.665...$

So by the time we're using four terms, the answer is the same as $2^{\sqrt{2}}$ calculated with a calculator out to three decimal digits. Notice that I've used the decimal representations of $\frac{4}{10} = 0.4$, $\frac{1}{100} = 0.01$, and $\frac{4}{1000} = 0.004$ here.

CHAPTER 5: POLYNOMIALS, FUNCTIONS, AND BEYOND

Pop Quiz: *Combining Like Terms of Polynomials*   (page 268)

1. $x + 1 + 3x^2 + 3x + 4$

   $3x^2 + (x + 3x) + (1 + 4) = 3x^2 + 4x + 5$

2. $x \bullet (2x - x^2) + x^3$

   $2x^2 - x^3 + x^3 = 2x^2 + (-x^3 + x^3) = 2x^2$

3. $x^4 + 3(x^2 - x^4 + 1) - 2$

   $x^4 + 3x^2 - 3x^4 + 3 - 2 = (x^4 - 3x^4) + 3x^2 + (3 - 2) = -2x^4 + 3x^2 + 1$

Pop Quiz: *Is That a Polynomial?*   (page 269)

If you said that the three expressions on the right are polynomials and the three on the left are not, then you're correct. But why?

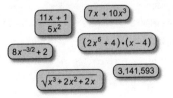

Well, the top left expression is not a polynomial because it has x^2 in its denominator, and variables with negative exponents are not allowed in polynomials. The center left expression has two problems: x has a negative exponent, and the exponent is not an integer. The bottom left expression is not a polynomial because of the square root. If the square root were absent, the expression would be a polynomial. How about the middle expression on the right? While the parentheses make it look a bit different than the other polynomials, if you use FOIL to multiply it out you'll see that it is a polynomial.

Pop Quiz: *Classifying Polynomials* (page 271)

- The product of a degree 1 monomial and a degree 2 trinomial is a . . .
  trinomial with a degree of __3__ .
- The sum of a degree 2 monomial and a degree 1 monomial is a . . .
  binomial with a degree of __2__ .
- The product of a degree 3 monomial and a degree 1 monomial is a . . .
  monomial with a degree of __4__ .
- The sum of a degree 2 monomial and a degree 1 binomial is a . . .
  trinomial with a degree of __2__ .

Pop Quiz: *Standard Form for Polynomials* (page 273)

1. $2 + 5x^2 - 3x = 5x^2 - 3x + 2$
2. $x^4 - 10 + x^3 + 5 = x^4 + x^3 - 5$ (after combining the constant terms)
3. $x + x^2 + 1 = x^2 + x + 1$
4. $x^2 - x + 2x^2 = 3x^2 - x$ (after combining the x^2 terms)

Pop Quiz: *Evaluating Polynomials* (page 277)

1. $5x^2 - 3x + 2$ at $x = 3$ is equal to $\rightarrow$ $5 \bullet 3^2 - 3 \bullet 3 + 2 = 45 - 9 + 2 = 38$
2. $x^4 + x^3 - 5$ at $x = 2$ is equal to $\rightarrow$ $2^4 + 2^3 - 5 = 16 + 8 - 5 = 19$
3. $x^2 + x + 1$ at $x = 4$ is equal to $\rightarrow$ $4^2 + 4 + 1 = 16 + 4 + 1 = 21$
4. $3x^2 - x$ at $x = 5$ is equal to $\rightarrow$ $3 \bullet 5^2 - 5 = 3 \bullet 25 - 5 = 75 - 5 = 70$

Pop Quiz: *Is That a Function?* (page 285)

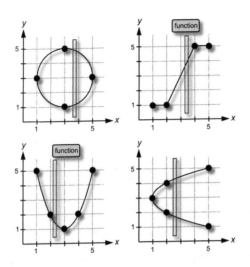

Pop Quiz: *What's the Domain and Range?* (page 287)

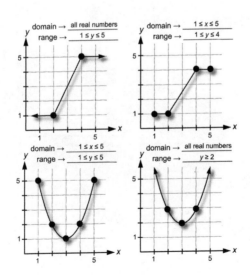

Pop Quiz: *Find the Slope of a Line*   (page 294)

Let's start by choosing the pair of points on the line (0, 1) and (2, 4). We can calculate the slope of this line using the equation

$$m = (y_2 - y_1) / (x_2 - x_1)$$

So, for this pair of coordinates, we find a slope of

$$m = (4 - 1) / (2 - 0) = 3/2$$

If we instead choose the coordinates $(-4, -5)$ and (0, 1), we get a slope of

$$m = [\, 1 - (-5) \,] / [\, 0 - (-4) \,] = (1 + 5) / 4 = 6/4 = 3/2$$

Yes, both pairs of coordinates give the same slope.

Pop Quiz: *Plot the Polynomial*   (page 300)

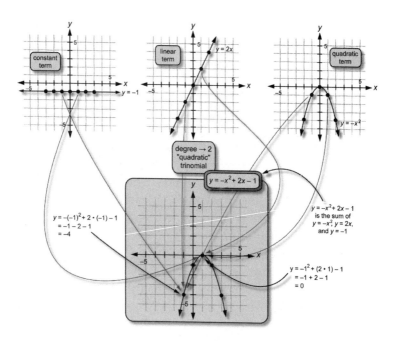

Pop Quiz: *Find the Equation of a Line . . . Then Plot It!*   (page 310)

As we found in the tutorial on How to Find and Write the Equations of Lines, once we know the slope of a line and the coordinates of one point on it, we can calculate the y-intercept using

$$b = y_1 - mx_1$$

Using the slope of $\frac{1}{2}$ and the (x_1, y_1) coordinates of the point (3, 2), we can find that

$$b = 2 - (\tfrac{1}{2} \cdot 3) = 2 - 3/2 = \tfrac{1}{2}$$

So, the equation for this line in slope-intercept form is

$$y = \tfrac{1}{2}x + \tfrac{1}{2}$$

and its graph looks like

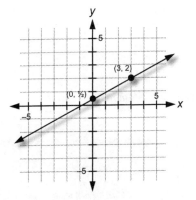

By the way, you don't need to tabulate any points to make this graph. Since you already know that the point (3, 2) is on the line, once you find the y-intercept you can simply connect these two points to construct the line.

Secret Agent Math-Lib: *What's That Dude Saying?*   (page 311)

First of all, it will be easier to solve these problems if you know which function is which in the plot. The function shown in black is $x^2 + 3x - 1$, and the one shown in gray is $-x^2 - 2x + 7$.

1. x^2+3x-1 .... <u>**0**</u> → <u>**0**</u>
2. $-x^2-2x+7$ .... <u>**E**</u> → <u>**12**</u>
3. x^2+3x-1 .... <u>**0**</u> → <u>**0**</u>
4. $-x^2-2x+7$ .... <u>**0**</u> → <u>**0**</u>
5. x^2+3x-1 .... <u>**P**</u> → <u>**0**</u>
6. $-x^2-2x+7$ .... <u>**N**</u> → <u>**0**</u>
7. x^2+3x-1 .... <u>**N**</u> → <u>**0**</u>
8. $-x^2-2x+7$ .... <u>**P**</u> → <u>**0**</u>
9. x^2+3x-1 to $3x+x^2-1$ .... <u>**A**</u> → <u>**7**</u>
10. $-x^2-2x+7$ to $-x^2-x \bullet 2+7$ .... <u>**M**</u> → <u>**21**</u>

So, what's that dude saying?

G O   F I G H T   W I N !

Is decrypted to say:

$\dfrac{N}{(9)} \dfrac{0}{(5)} \qquad \dfrac{F}{(1)} \dfrac{I}{(8)} \dfrac{G}{(3)} \dfrac{H}{(7)} \dfrac{T}{(4)}. \qquad \dfrac{R}{(10)} \dfrac{U}{(2)} \dfrac{N}{(6)} !$

Pop Quiz: *Let's Solve a System of Equations!*   (page 328)

Our job is to solve the following system of linear equations using the method of elimination:

$$2x+3y+z=1 \qquad\qquad -x-y+2z=2 \qquad\qquad x+2y-2z=3$$

Let's begin by multiplying both sides of the last equation by -1 with the goal of eliminating x from the system of equation. This gives us:

$$
\begin{array}{r}
2x+3y+z = 1 \\
-x-y+2z = 2 \\
-x-2y+2z = -3 \\
+ \hline
0x+0y+5z = 0
\end{array}
$$

Wow! A lot of stuff canceled out for us there. So what does this tell us? Well, we can now solve and find that $z=0$. Now that we have one variable, let's plug this back into two of the three original equations (we only need two since we're now only solving for two variables):

$$-x-y=2$$
$$x+2y=3$$
$$+\underline{\hspace{3cm}}$$
$$y=5$$

By choosing to use the second and third equations, I was able to quickly solve for $y=5$ without having to do any multiplication. Now that we have $y=5$ and $z=0$, we can plug these values into any of the three original equations to find the value of x. Let's use the second equation:

$$-x-5=2$$

Which gives us $x=-7$. So the full solution to the problem is $x=-7$, $y=5$, and $z=0$. You should try plugging this solution into the original equations and make sure it works.

Pop Quiz: *Let's Solve a System of Inequalities!*   (page 332)

1. $y>-x+3$, $y<x+1$, and $x\leq 4$
2. $y\geq -x+3$, $y\leq x+1$, and $y>2$
3. $y\geq -x+3$, $y<x+1$, and $x<-2$

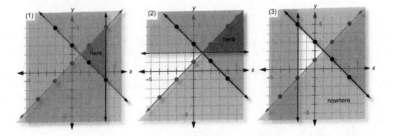

The two systems on the left have solutions where the shaded regions from all three inequalities overlap (marked with the word "here"). How about the system on the right? Well, there's nowhere that all three regions overlap, so this system of linear inequalities has no solution.

Final Exam   (page 335)

• Polynomials

1.

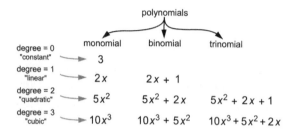

• Functions

2. On the left, you can see the polynomial $y = \frac{x^4}{4}$ plotted on the same set of axes as the parabola $y = \frac{x^2}{4}$. Both functions look similar, but the degree 4 function increases much faster as you go out in the positive or negative x direction. On the right you can see the degree 5 polynomial $y = \frac{x^5}{4}$ next to the cubic $y = \frac{x^3}{4}$. Again, both look similar, but the higher degree polynomial increases much faster.

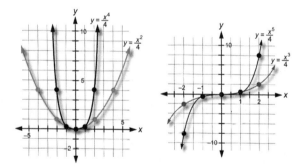

Notice that the two functions on the left have even exponents while the exponents of the two functions on the right are odd. Even powered polynomials are always symmetric around the y-axis (including the constant polynomial $y = x^0$), and are said to have "even" symmetry. Odd powered polynomials are always symmetric around the line $y = x$ (including the line $y = x$ itself), and are said to have "odd' symmetry.

• Plotting Lines and Other Polynomials

3. As we saw in the chapter, the $-x$ term in the polynomial $y = x^2 - x - 4$ causes the resulting parabola to be shifted to the *right* side of the y-axis. In the case of the polynomial $y = x^2 + x - 2$, the linear term is now $+x$ which you can see causes the parabola to shift to the *left* of the y-axis.

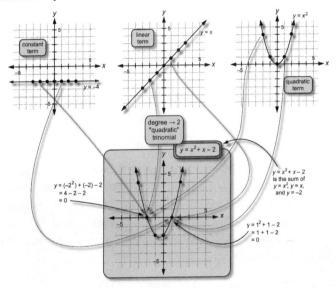

• Systems of Equations

4. Solve the following system of equations by substitution . . .

$2x + y = 5$ and $4x + 2y = 1$

Let's start by solving the equation on the left for y. The answer is

$y = -2x + 5$

Now let's plug this into the equation on the right:

$4x + 2 \bullet (-2x + 5) = 1$

If we simplify this, we get

$4x - 4x + 10 = 1$

Which says that $10 = 1$! Well that's definitely not right . . . right? I warned you that something "funny" was going to happen, so what is it? Take a look at the graph of the two equations:

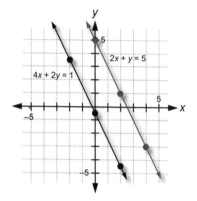

Aha! Well, there's your problem. Those two lines are parallel! And the whole point of solving a system of linear equations is to figure out where the two lines intersect. But parallel lines don't intersect, so you end up getting a nonsense answer—which should always be a clue to investigate a little further. As we talked about in the chapter, parallel lines have the same slope. And if you put both of the equations for these lines in slope-intercept form, you'll find that they both have a slope of -2.

• Systems of Inequalities

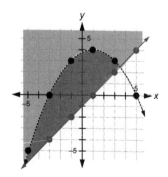

5. In our bizarre monkey story, monkeys have free reign to roam everywhere above the dashed line (but not on it) as determined by the inequality $y > -\frac{1}{4}(x-1)^2 + 4$, and you have free reign to roam everywhere above the solid line (including on it) as determined by the ine-

quality $y \geq x - 1$. The light-shaded region shows the part of the x-y plane that is a solution to the system of inequalities. In other words, this is the region where both you and the monkeys can mingle. How about the dark-shaded region? Well, that's your "safety zone." Not that howler monkeys are dangerous, but if you were worried for some reason, then this dark-shaded region shows you where to hang out!

CHAPTER 6: THE ROOT OF THE PROBLEM

Pop Quiz: *On a More General Note . . .* (page 342)

The equations we need to write to solve this problem are very similar to the ones we wrote for the second challenge problem. First of all, the problem asks us to write an equation saying that the product of two numbers is equal to P, which is just

$$n_1 \bullet n_2 = P$$

The only difference between this equation and the one we wrote before is that we've switched P for the constant 1056.

The second equation needs to express the relationship between the two variables n_1 and n_2. Instead of these numbers being consecutive (which means they differ by one), here we're allowing them to be separated by any integer amount, S. We can write this like

$$n_2 = n_1 + S$$

The only difference between this equation and the one we wrote before is that we've switched S for the constant 1.

Now we need to put these two equations together to come up with an equation for n_1. Let's do that by substituting the second equation into the first:

$$n_1 \bullet (n_1 + S) = P$$

Or, if we multiply this out, we get

$$n_1^2 + Sn_1 = P$$

As we'll learn about later in the chapter, we can subtract P from both sides to get an equation in the general form of a quadratic equation:

$$n_1^2 + Sn_1 - P = 0$$

If we set $S = 1$ (representing a separation of 1) and $P = 1056$, we get back the equation we found in the challenge problem.

Pop Quiz: *Where Are the Roots?*   (page 346)

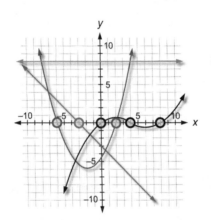

- The constant polynomial has __0__ roots.
- The linear polynomial has __1__ root located at $x =$ __–3__.
- The quadratic polynomial has __2__ roots located at $x =$ __–6__ and __2__.
- The cubic polynomial has __3__ roots located at $x =$ __0__, __4__, and __8__.

- Does the root of the linear polynomial solve the equation $-x - 3 = 0$?

    Root 1: $x = -3 \rightarrow -(-3) - 3 = 3 - 3 = 0$ ✓

- Do the roots of the quadratic polynomial solve the equation $x^2 + 4x - 12 = 0$?

    Root 1: $x = -6 \rightarrow (-6)^2 + 4(-6) - 12 = 36 - 24 - 12 = 0$ ✓
    Root 2: $x = 2 \rightarrow 2^2 + 4 \bullet 2 - 12 = 4 + 8 - 12 = 0$ ✓

- Do the roots of the cubic polynomial solve the equation $x^3 - 12x^2 + 32x = 0$?

Root 1: $x = 0 \rightarrow 0^3 - 12 \bullet 0^2 + 32 \bullet 0 = 0$ ✓
Root 2: $x = 4 \rightarrow 4^3 - 12 \bullet 4^2 + 32 \bullet 4 = 64 - 192 + 128 = 0$ ✓
Root 3: $x = 8 \rightarrow 8^3 - 12 \bullet 8^2 + 32 \bullet 8 = 512 - 768 + 256 = 0$ ✓

Pop Quiz: *How Many Roots?*   (page 349)

To determine the number of roots a polynomial has, we first have to fully multiply it out using the distributive property and FOIL so that we can see each of its terms. The degree of the polynomial is the highest power of the variable in all the terms. And the number of roots is equal to the degree of the polynomial.

Roots

1. $x^2 \bullet (x+1) = x^3 + x^2 \rightarrow$ degree $= 3$ .... __3__
2. $x^3 - 2x^2 + x + 3 \rightarrow$ degree $= 3$ .... __3__
3. $(x^2 + x) \bullet x^2 = x^4 + x^3 \rightarrow$ degree $= 4$ .... __4__
4. $(x+1) \bullet (x+1) = x^2 + x + x + 1 = x^2 + 2x + 1 \rightarrow$ degree $= 2$ .... __2__

Pop Quiz: *Reducible or Irreducible?*   (page 353)

The reducible polynomials are: B, C, and D (since they all cross the x-axis).
The irreducible polynomials are A and E (since they do *not* cross the x-axis).

Pop Quiz: *What's the Multiplicity of Each Root?*   (page 355)

1. $x^2 - 4x + 4 = (x-2) \bullet (x-2)$

    Root 1: Located at $x =$ __2__ with multiplicity __2__ .

2. $x^3 - 2x^2 + x = (x-0) \bullet (x-1) \bullet (x-1)$

    Root 1: Located at $x =$ __0__ with multiplicity __1__ .
    Root 2: Located at $x =$ __1__ with multiplicity __2__ .

3. $x^3 - x^2 - x + 1 = (x-1) \bullet (x+1) \bullet (x-1)$

Root 1: Located at $x=$ __1__ with multiplicity __2__.
Root 2: Located at $x=$ __−1__ with multiplicity __1__.

Let's take a look at the last problem in a little more detail. The equation that we're trying to solve is

$$x^3 - x^2 - x + 1 = 0$$

If we use the equivalent form of the polynomial shown in the problem, we can write this as

$$(x-1) \bullet (x+1) \bullet (x-1) = 0$$

You can multiply out everything on the left side of this expression using FOIL and the distributive property to check and see that the two expressions for the polynomial really are equivalent, if you like. But the important thing to notice about this last equation is that it has two solutions: $x=1$ or $x=-1$. Plugging either of those values in for x will make the equation true. But there are also two different places where you could plug $x=1$ in, which means that the root $x=1$ has a multiplicity of 2.

Pop Quiz: *Picturing Polynomials in Factored Form*   (page 359)

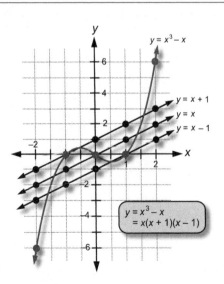

As you can see, the polynomial produced by multiplying the linear polynomials x, $x+1$, and $x-1$ is the cubic polynomial $x^3 - x$.

Pop Quiz: *Find the Prime Factors* (page 362)

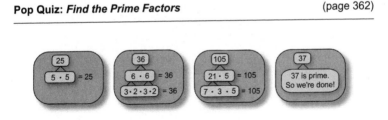

Pop Quiz: *Find the Greatest Common Factor of a Group of Numbers* (page 364)

To find the greatest common factor, you first need to do the factorization by filling out the factorization circles. Actually, to find just the greatest common factor, you only need to complete the inner ring of the factorization circles. But I've done the whole thing here . . . just for fun.

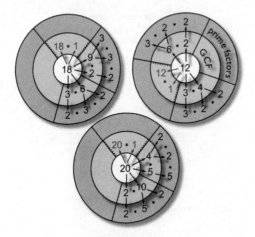

Now all you have to do is find the largest number in common from the first ring of all three factorization circles. You should start with the biggest number and check to see if it exists in all the other first rings. If it doesn't, move on to the next largest number. The greatest common factor here is 2—which isn't so "great." But that's how it goes sometimes.

Pop Quiz: *Find the Greatest Common Polynomial Factor* (page 366)

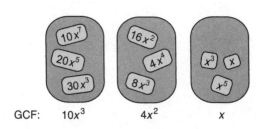

GCF: $10x^3$ $4x^2$ x

Algebra Tutorial: *How to Factor Polynomials* (page 372)

Step 1: Factor Out the Greatest Common Factor

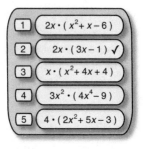

1 $2x \cdot (x^2 + x - 6)$

2 $2x \cdot (3x - 1)$ ✓

3 $x \cdot (x^2 + 4x + 4)$

4 $3x^2 \cdot (4x^4 - 9)$

5 $4 \cdot (2x^2 + 5x - 3)$

Step 2: Factor Using Identities

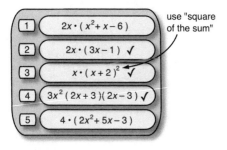

1 $2x \cdot (x^2 + x - 6)$

2 $2x \cdot (3x - 1)$ ✓

3 $x \cdot (x + 2)^2$ ✓

4 $3x^2 (2x + 3)(2x - 3)$ ✓

5 $4 \cdot (2x^2 + 5x - 3)$

use "square of the sum"

Step 3: Factor Using Identities

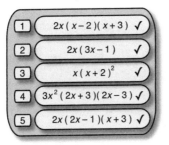

1	$2x(x-2)(x+3)$ ✓
2	$2x(3x-1)$ ✓
3	$x(x+2)^2$ ✓
4	$3x^2(2x+3)(2x-3)$ ✓
5	$2x(2x-1)(x+3)$ ✓

Pop Quiz: *Factor by Grouping*   (page 383)

The polynomial we want to factor is:

$$x^3 - x^2 - 4x + 4$$

As suggested in the problem, let's try factoring this problem by grouping the first two terms and the last two terms, and finding separate common factors for each:

$$(x^3 - x^2) - (4x - 4)$$

Notice that I had to switch the signs around in the last term since I put the negative sign on the outside of the parentheses. In the first term, the greatest common factor is x^2, and in the second term the greatest common factor is 4. When we factor those out, we get

$$x^2(x-1) - 4(x-1)$$

And now you see the magic that can sometimes happen when you factor by grouping: a common factor of $x-1$ has appeared in both terms! Let's factor out $x-1$ to get

$$(x-1)(x^2-4)$$

Notice that the factor on the right is actually the difference of squares—which is more obvious if we write it $x^2 - 2^2$. Using the identity for the difference of squares, we get a final factorization of

$$(x-1)(x+2)(x-2)$$

Pop Quiz: *Standard Form* (page 384)

1. $(2x+5)(x-2)=0 \rightarrow 2x^2-4x+5x-10=2x^2+x-10$

   Standard form $\rightarrow$ __2__ x^2+ __1__ $x+$ __-10__ $=0$

2. $(x-3)^2+1=0 \rightarrow (x-3)(x-3)+1=x^2-3x-3x+9+1=x^2-6x+10$

   Standard form $\rightarrow$ __1__ x^2+ __-6__ $x+$ __10__ $=0$

3. $(x+1)(x-1)+1=0 \rightarrow x^2-x+x-1+1=x^2$

   Standard form $\rightarrow$ __1__ x^2+ __0__ $x+$ __0__ $=0$

Pop Quiz: *Quadratic Equations with "Missing" Terms* (page 386)

- **No linear term:** Plug the solutions $x=\pm[-c/a]^{1/2}$ into the equation $ax^2+c=0$ and verify that the solution works.

   1. $a([-c/a]^{1/2})^2+c=a\bullet(-c/a)+c=-c+c=0$ ✓
   2. $a(-[-c/a]^{1/2})^2+c=a\bullet(-c/a)+c=-c+c=0$ ✓

- **No constant term:** Plug the solutions $x=0$ and $x=-b/a$ into the equation $ax^2+bx=0$ and verify that the solution works.

   1. The solution for $x=0$ is pretty simple and quickly gives $0=0$ ✓
   2. $a(-b/a)^2+b(-b/a)=a\bullet b^2/a^2-b^2/a=b^2/a-b^2/a=0$ ✓

Pop Quiz: *Let's Solve Equations with the Quadratic Formula!* (page 388)

The problem asks us to use the quadratic formula to solve

$$x^2+7x-918=0$$

This quadratic has $a=1$, $b=7$, and $c=-912$. Let's now use these coefficients in the quadratic formula to find the roots of the equation.

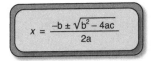

First, let's calculate the portion of the formula inside the square root

$$b^2 - 4ac = 7^2 - (4 \bullet 1 \bullet -918) = 49 + 3672 = 3721$$

Since this number is positive, the equation has valid solutions. Now let's use this number and find both roots:

$$\text{Root 1} \rightarrow x = (-7 + \sqrt{3721})/2 = 54 / 2 = 27$$
$$\text{Root 2} \rightarrow x = (-7 - \sqrt{3721})/2 = -68 / 2 = -34$$

Secret Agent Math-Lib: *Will Secret Agent Math Escape this Nightmare?* (page 389)

The coefficients of the quadratic equation in the problem are:

$$a = \underline{\;\;1\;\;} \quad b = \underline{\;\;1\;\;} \quad c = \underline{\;-1056\;}$$

and the roots of the quadratic equation with these coefficients are therefore given by:

$$n_1 = \frac{-\underset{(b)}{1} \pm \sqrt{\left(\underset{(b)}{1}\right)^2 - 4 \cdot \underset{(a)}{1} \cdot \underset{(c)}{-1056}}}{2 \cdot \underset{(a)}{1}}$$

This equation can be simplified to:

$$n_1 = -\frac{1}{2} \pm \frac{1}{2} \sqrt{1 - 4 \cdot \underset{(c)}{-1056}}$$

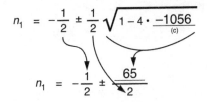

$$n_1 = -\frac{1}{2} \pm \frac{65}{2}$$

Since the value of the expression inside the radical is a positive number, this quadratic equation has valid solutions. The two solutions for n_1 are:

plus!

$$\text{solution 1:} \quad n_1 = -\frac{1}{2} + \frac{65}{2} = \frac{64}{2} = \underline{32}$$

$$\text{solution 2:} \quad n_1 = -\frac{1}{2} - \frac{65}{2} = \frac{-66}{2} = \underline{-33}$$

minus!

and the corresponding solutions for n_2:

$$\text{solution 1:} \quad n_2 = \frac{32}{(n_1)} + 1 = \underline{33}$$

$$\text{solution 2:} \quad n_2 = \frac{-33}{(n_1)} + 1 = \underline{-32}$$

That's it! We now have our two complete solutions:

$$\text{solution 1:} \quad n_1 = \underline{32} \qquad n_2 = \underline{33}$$
$$\text{solution 2:} \quad n_1 = \underline{-33} \qquad n_2 = \underline{-32}$$

Check and make sure both of these solutions work! $32 \cdot 33 = 1056$ and $-33 \cdot -32 = 1056$. So yes, they both work. Algebra is pretty cool, right?

Final Exam   (page 410)

• Roots of Polynomials

1. Here are the equations for both the quadratic and cubic polynomials shown in the graph.

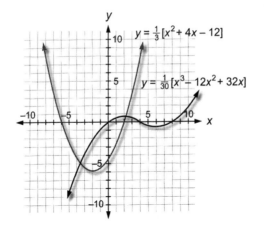

The quadratic polynomial has two roots located at $x=-6$ and $x=2$. Therefore, we can write an equation for it like this:

"quadratic" $= c \cdot (x+6) \cdot (x-2)$

If we multiply this out, we get

"quadratic" $= c \cdot [x^2 + 4x - 12]$

Now let's evaluate this at $x=3$ to see if we need a constant. At $x=3$, the value is $3^2 + 4 \cdot 3 - 12 = 9 + 12 - 12 = 9$. As you can see from the graph, the actual polynomial has a value of 3 at $x=3$. Therefore, we need to set the constant equal to $\frac{1}{3}$. The final equation is therefore

"quadratic" $= \frac{1}{3}[x^2 + 4x - 12]$

If you plug $x=3$ into this equation, you get the correct answer of 3. Similarly, for the cubic polynomial, the three roots are located at $x=0$, $x=4$, and $x=8$. Therefore, we can write an equation like this:

"cubic" $= c \cdot x \cdot (x-4) \cdot (x-8)$

Once we multiply this out, plug in a number to find the required constant, and put it all together, we get

"cubic" $= (1/30) \cdot [x^3 - 12x^2 + 32x]$

• Factoring Polynomials

2. After using FOIL to multiply out $(x+y) \cdot (x-y)$, the meaning of the resulting identity becomes much easier to understand. Just follow the arrows below, and make sure you check whether each part is being added or subtracted!

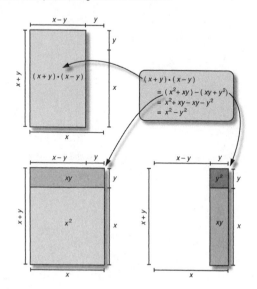

• Solving Quadratic Equations

3. Since g is in the denominator of the fraction that determines how long it takes the rock to fall

$$t_{\text{down}} = \sqrt{2d/g}$$

and since we're pretty sure that the rock would fall slower on the moon than on earth, it must be the case that little g is smaller on the moon than on earth. Why? Because If g is smaller, than a smaller number in the denominator means that the fraction itself gets bigger . . . which means the time for the rock to fall gets bigger . . . just as we think it should. And that's exactly right. Since the mass of the moon is so much less than the mass of earth, it has a much weaker gravitational pull on you or a rock or anything else. So things fall slower, people bounce, and the number "little g" is smaller.

Acknowledgments

I'm willing to wager that among the top ten things acknowledged by authors in sections like this one is an expression of disbelief at just how all-consuming and, frankly, tough it is to finish a book. And now I know exactly why that is—because this book was a ton of work! But I honestly enjoyed every minute of the process (well, okay . . . almost every minute). Upon hearing the news that I had finished the final draft, an astronomy colleague declared: "Congratulations! You're now the proud father of a bouncing baby book." I don't have any kids of my own, but I think that's a pretty good description of how I felt when it was done.

Of course, the fact that the book was ever finished at all is due to the hard work of many people besides myself. I'd be remiss not to thank the folks at Macmillan/Quick and Dirty Tips for all their support and for giving me the opportunity to become their Math Dude. In particular, I'd like to thank Mignon Fogarty and Richard Rhorer for getting this ship built and for navigating it so capably. Thanks also to the entire St. Martin's production staff for dealing with all the complexities presented by this book and for making the end product look so great.

It's no exaggeration to say that there's one person in particular without whom I could not have finished this book, and that's Emily Rothschild. Her keen eye and insight, as well as her seemingly infinite reservoir of dedication, kindness, and encouragement saw me through the difficult patches and helped me find my way back home. Thank you, Emily.

While it's recommended to maintain sanity at all times, it's especially important when you're trying to put together a coherent and cohesive guide to a complex subject like algebra. So I owe a huge debt of gratitude to all the family and friends that reeled me in when I threatened to cross the tenuous spider-web of a line between slightly nuts and full-blown fall-off-the-rocker crazy. In particular, I'm forever grateful to my wife, Shannon, for taking care of (and putting up with) me during those last few months of writing.

Finally, thanks to all the listeners and supporters of the Math Dude podcast. And a special thanks to all my Facebook and Twitter fans and followers. It's a ton of fun for me to get to share my knowledge with you every day, but it's even more fun to get to learn so much from you in return.

Thank you all!

INDEX

absolute value equations:
  exam questions on, 195
  picturing, 187–91
  solving graphically, 178–79
absolute values, 144–50, 155
  calculating distance between
    numbers, 148–50
  calculating distance between
    variables, 153–54
  exam questions on, 194
  linear equations with, 174–79
  notation for, 145
  of numbers, calculating, 145
  of variables, calculating, 145–47
  writing distances on the number
    line, 154–55
addition, 11, 198
  associative property of, 224–25
    exam questions on, 261
  commutative property of, 223,
    225
    exam questions on, 261
  of fractions, 202–4
  in golden rule of equation
    solving, 108
  of like terms, 235–36
  in moving variable to one side
    of equation, 112–13, 169–70
  of negative numbers, 211, 212–14

in order of operations, 90–91
  of polynomials, 297
  quick, 224–25
  repeated, 205–6
  with roots, 252–53
  symbol for, 12
  of variables, 214–15
algebra:
  basics of, 76–132
  fundamental theorem of, 348–49,
    356–57
  life made easier by, 40–41
  picturing the meaning of,
    100–104
  point of, 259
  in real world, 33–75, 259–60
    problem solving, 123–29
Algebra Decoder, 6
  absolute value, 145
  area, 200
  digits, 140
  inverse process, 199
  letter x, 81, 82, 83
  mean and range, 166–67
  plus or minus, 114
  prime number, 361
  Pythagorean, 51
  quadratic equation, 343
  right angle, 43–44

Algebra Decoder (*cont'd*)
    term, 215
    volume, 218
algebraic expressions,
    see expressions
algebra tutorials, 4–5
    combining like terms, 238–39
    equations of lines, 300–310
    factoring polynomials, 372–82
    graphs, 63–71
    order of operations, 88–93
    simplifying expressions, 220–23
    solving an equation, 107–15
    solving linear equations with
        absolute values, 176–78
    solving linear inequalities,
        181–86
    solving real-world algebra
        problems, 123–29
    solving single variable linear
        equations, 167–72
    solving system of equations,
        321–27
    solving system of inequalities,
        329–32
angle, right, 43–44
area, 200
arithmetic, 77, 82, 197–262
    basic, 11–12
    combining like terms in,
        234–40
    Egyptian, 137–40
    exponentiation in,
        see exponentiation
    preparing for algebra, 198–201
    properties of, 223–31
        exam questions on, 261
    with "real" numbers, 201–13
    symbols used in, 12, 77–78, 81
        exam questions on, 28–29
associative properties:
    of addition, 224–25
        exam questions on, 261
    of multiplication, 226

base, 19, 241
binomials, 269–70
brackets and braces, 92
budget, 41

Cartesian coordinates, 64–65
challenge problems, 71–72, 333–35,
    339–46, 389–93, 401–9
combining like terms, 234–40
commutative properties:
    of addition, 223, 225
        exam questions on, 261
    of multiplication, 225
complex numbers, 162

constants, 79–80
    pi (π), 80–81, 161
coordinate plane, 65
    plotting points on, 67–68
cubing numbers, 241, 242, 357

decimal numbers, 140–43, 158–59
    irrational, 160–62
        exponents, 256–57
deck problem, 58–70
denominator, 158, 159
    common, 202–3
Descartes, René, 64
digits, 140
distance:
    calculating between numbers,
        148–50
    calculating between variables,
        153–54
    writing on number line, 154–55
distributive property, 226–31,
    367–68
    exam questions on, 261
division, 11, 201
    with exponents, 244–46
    of fractions, 207
    in golden rule of equation
        solving, 109
    of inequalities by negative
        numbers, 184
    in isolating variable in equations,
        113, 170–71
    of negative numbers, 210, 212–14
    in order of operations, 90–91
    of polynomials, 299
    with roots, 252
    symbol for, 12
    of variables, 219
    by zero, 233–34
dog toy problem, 123–29, 132

Egyptian arithmetic, 137–40
equations, 40, 95–100, 106–17
    challenge problems, 71–72,
        333–35, 339–46, 389–93,
        401–9
    checking solution of, 115, 171
    exam questions on, 130–32
    functions and, 279–80, 283,
        288–89
    geometric meaning of, 103–4
    golden rule of solving, 107–10
    isolating variable in, 113, 170–71
    linear, *see* linear equations
    of lines, 300–310
        horizontal and vertical, 309–10
    moving variable to one side of,
        112–13, 169–70
    number of unknowns in, 320–21

order of operations in, 88–93
  exam questions on, 131
purpose of, 96
quadratic, 342–44
  exam questions on, 411–12
  factoring and, 387
  quadratic formula, 387–88
  solving, 383–89
  standard form of, 383–84
  visualizing solution of, 343–44
  with all terms, 386
  with no constant term, 385, 386
  with no linear term, 384–85,
    386
scale metaphor for, 106–7, 279–80
simplifying each side of, 111–12,
  168–69
solving graphically, 116–17
systems of, 320–28
  exam questions on, 337
  solving, 321–27
translating English to, 166–67
writing and solving, 96–98,
  107–15
even numbers, 10, 11
exponentiation, 240–57
  base in, 19, 241
  cubes, 241, 242, 357
  exam questions on, 261–62
  picturing, 241–42
  roots and, 250–51, 254–56
  see also square roots; squaring
    numbers
exponents, 19–20, 241
  division with, 244–46
  with exponents, 247
  fractional, 249–50, 254–55
  in golden rule of equation
    solving, 109–10
  in isolating variable in equations,
    113
  multiplication with, 242–44
  in order of operations, 89–91
  summary of properties of, 248
  zero as, 246–47
expressions, 86–88, 95, 96, 97, 99,
  100, 274
  absolute value of, 145, 147
  examples of, 86–87
  exam questions on, 130
  geometric meaning of, 101–3
  rational, 273–74
  simplifying, 220–23
  translating English to, 87–88

factored form, writing polynomials
  in, 356–59
factoring:
  numbers, 360–62, 367

polynomials, 360–83
  exam questions on, 411
  solving quadratic equation by, 387
factors:
  greatest common:
    of a group of numbers, 362–64
    of a group of polynomials,
      364–66
  prime, 360–62, 363
final exams, 5, 27–30, 72–75,
  130–32, 194–96, 260–62,
  335–38, 410–12
FOIL, 230–33, 367–68
fractions, 157–59, 163, 202–8
  adding and subtracting, 202–4
  denominator in, 158, 159
    common, 202–4
  dividing, 207
  exam questions on, 260–61
  as exponents, 249–50, 254–55
  multiplying, 204–6
  numerator in, 158
functions, 279–96
  constant, on graphs, 289–90
  domain and range of, 285–87
  equations and, 279–80, 283,
    288–89
  exam questions on, 336
  how they work, 280–81
  how to picture, 281–83
  linear, on graphs, 290–91
  polynomial, visualizing, 287–91,
    294–95
  quadratic and cubic, on graphs,
    294–96
  slope in, 291–92
    calculating, 292–94

geometric meaning of equations,
  103–4
geometric meaning of expressions,
  101–3
graphs:
  Cartesian coordinates in, 64–65
  connecting dots to draw a
    curve on, 69–71
  exam questions on, 74–75
  functions on:
    constant, 289–90
    linear, 290–91
    quadratic and cubic, 294–96
  making, 63–71
  ordered pairs in, 65–71
  plotting points on, 67–68
  solving absolute value equations
    with, 178–79
  solving equations with, 116–17
  solving linear equations with,
    172–73

greater than, 37, 57
greater than or equal to, 38
greatest common factor (GCF):
    of a group of numbers, 362–64
    of a group of polynomials, 364–66
grocery shopping problem, 34–37,
    38–40

hypotenuse, 49–50, 65–66, 74, 161

imaginary numbers, 162
    polynomial roots, 351–53
inequalities, 37–38, 57
    exam questions on, 73, 195–96
    interval notation and, 186–87
    linear, 179–91
        solving, 181–86
    multiplying and dividing by
        negative numbers, 184
    picturing, 187–91
    systems of, 328–33
        exam questions on, 337–38
        solving, 329–32
infinity, 143, 155, 157
integers, 156–57, 158–60, 162, 163
interval notation, 186–87
inverse process, 199
irrational numbers, 160–61, 162,
    163
    exponents, 256–57
    pi (π), 80–81, 161
    square root of 2, 161–62

Knot Dude, 42–46, 50, 51, 52, 56

less than, 37, 57
less than or equal to, 37
linear equations, 164–75
    with absolute values, 174–79
    exam questions on, 194
    single variable, 165–66
        solving, 167–74, 176–78
        solving graphically, 172–73
    with two variables, 191–93
linear inequalities, 179–91
    solving, 181–86
lines:
    equations of, 300–310
        horizontal and vertical, 309–10
    exam questions on, 336–37
    parallel and perpendicular,
        slopes of, 318–20

magnitude, 144, 145, 155, 204
math brain games, 5–6
    deck problem, 58–62
    distributive property, 229–31
    monkey and banana problem,
        314–18, 320

picturing absolute value equations
    and inequalities, 187–91
picturing the meaning of algebra,
    100–104
polynomial identity, 368–71
positively odd puzzle, 23–26
mathematical constants, 79–80
    pi (π), 80–81, 161
mean, 166–67
monkey and banana problem,
    314–18, 320
monomials, 269–70
multiplication, 11, 199–200, 240
    associative property of, 226
        exam questions on, 261
    commutative property of, 225
        exam questions on, 261
    with exponents, 242–44; see also
        exponentiation; exponents
    factoring and, 360–62, 367
    FOIL formula and, 230–33,
        367–68
    of fractions, 204–6
    in golden rule of equation
        solving, 109
    of inequalities by negative
        numbers, 184
    in isolating variable in equations,
        113, 170–71
    of negative numbers, 208–10,
        212–14
    in order of operations, 90–91
    of polynomials, 299
    quick, 231–33
    as repeated addition, 205–6
    with roots, 252
    symbol for, 12
    of variables, 216–19
multiplication table, 22

negative numbers, 10, 11, 155–56
    adding, 211, 212–14
    dividing, 210, 212–14
    multiplying, 208–10, 212–14
    multiplying and dividing
        inequalities by, 184
    PEMDAS and, 93
    subtracting, 212–14
number line, 143, 162, 163
    absolute values and, see absolute
        values
    exam questions on, 194
    fractions on, 157–58, 159
    integers on, 156–57
numbers, 78, 135–96
    absolute value of, see absolute
        values
    calculating distance between,
        148–50

complex, 162
decimal, *see* decimal numbers
even, 10, 11
exam questions on, 194
in expressions, 86–87, 97
factoring, *see* factoring; factors
fractions, *see* fractions
history of, 136–41
imaginary, 162
    polynomial roots, 351–53
infinity, 143, 155, 157
integers, 156–57, 158–60, 162, 163
irrational, 160–61, 162, 163
    exponents, 256–57
    square root of 2, 161–62
magnitude of, 144, 145, 155, 204
negative, 10, 11, 155–56
    adding, 211, 212–14
    dividing, 210, 212–14
    multiplying, 208–10, 212–14
    multiplying and dividing
        inequalities by, 184
    PEMDAS and, 93
    subtracting, 212–14
odd, *see* odd numbers
patterns in, 17–18, 20–26
    exam questions on, 29–30
positive, *see* positive numbers
prime, 360–62
properties of, 10–11
    exam questions on, 27–28
rational, 157–60, 162, 163
real, 162–63
"small," 144, 145
zero, 155, 156
    dividing by, 233–34
    as exponent, 246–47
numerals, 78–80, 81
numerator, 158

odd numbers, 10, 11
    positive, 10–11
        adding up, 9–10, 13–14, 17–26
operators, 86–87, 97, 100
ordered pairs, 65–71
order of operations, 88–93
    exam questions on, 131
overachiever badges, 6
    converting units, 399–400
    function vs. equation, 288–89
    interval notation, 186–87
    irrational exponents, 256–57
    linear equations with two
        variables, 191–93
    polynomials, 273–74
        imaginary roots of, 352–53
        multiplication and division, 299
        multivariate, 277–78

proving the Pythagorean
    theorem, 104–6
Pythagorean triples, 51–53
repeated addition, 205–6
solving absolute value equations
    graphically, 178–79
solving equations graphically,
    116–17
solving linear equations
    graphically, 172–73

parallel lines, slopes of, 318–20
parentheses, 89, 90–91, 92
patterns, in numbers, 17–18, 20–26
    exam questions on, 29–30
PEMDAS, 90–91, 93
    negative numbers and, 93
perfect squares, 18–26, 46
    exponents and, 19–20
perpendicular lines, slopes of,
    318–20
pi (π), 80–81, 161
plus or minus, 114
polynomials, 260, 265–78, 279
    adding, 297
    anatomy of, 268–69
    building, 266–67
    combining like terms in, 267–68
    cubic, 271
    degree of, 269–71, 348–50, 351,
        354
    division of, 299
    equations of lines, 300–310
    evaluating, 274–77
    exam questions on, 335, 336–37
    factoring, 360–83
        exam questions on, 411
        method for, 372–82
    functions, visualizing, 287–91,
        294–95
    and fundamental theorem
        of algebra, 348–49, 356–57
    greatest common factor of,
        364–66
    identity of, 368–71
    irreducible, 350–51, 353, 356
    linear, 270–71
    monomials, binomials, and
        trinomials, 269–70
    multiplication of, 299
    multivariate, 277–78
    plotting, 300
    quadratic, 271
    reason for using, 271–72
    reducible, 349–50, 353, 356
    roots of, 344–59
        exam questions on, 410–11
        imaginary, 351–53
        multiplicity of, 354–55

polynomials (*cont'd*)
  solving problems with, 296–99
  subtracting, 298–99
  writing in factored form,
    356–59
pop quizzes, 4
  absolute value equation, 179
  absolute values on the number
    line, 147–48
  adding and subtracting fractions,
    203
  adding quickly, 224–25
  algebra, decimal numbers,
    and addition, 142–43
  ancient arithmetic, 139–40
  budget, 41
  combining like terms, 240
    of polynomials, 268
  converting fractions to decimal
    numbers, 159
  division with exponents, 246
  dots vs. numerals, 79
  drawing a curve on a graph,
    70–71
  English to equation translation,
    166–67
  equality vs. inequality, 180–81
  equations of lines, 310
  equation solving, 116
  equation to find two numbers
    separated by S amount, whose
    product is P, 342
  exponents, 248
    even and odd, 242
    finding, 252
  expressions vs. equations, 99
  Four Fours, 93–94
  fractions, arithmetic, and algebra,
    207–8
  functions, 284–85
    domain and range of, 286–87
  greatest common factor:
    of a group of numbers, 364
    of a group of monomials, 366
  inequalities, 38
  inequality to interval notation
    translation, 187
  linear equations, 164
  multiplying and dividing by
    negative numbers, 210
  multiplying fractions, 206
  multiplying quickly, 231–33
  multiplying variables, 217
  multiplying with exponents, 244
  naming numbers, 11
  perfect squares, 20
  polynomials, 269
    classifying, 271
    evaluating, 277

factoring, 383
  picturing in factored form, 359
  plotting, 300
  reducible and irreducible, 353
  roots of, 346–48, 349, 355
  standard form for, 272–73
positive and negative outcomes,
    213–14
prime factors, 362
properties of arithmetic, 228–29
quadratic equations:
  standard form of, 384
  with missing terms, 386
quadratic formula, 388
right angles, 44
roots, 254
  combining powers and, 256
simplifying expressions, 223
single variable linear equation,
    173–74
slope of a line, 294
square roots, 49
system of equations, 328
system of inequalities, 332–33
translating English to algebraic
    expressions, 87–88
types of numbers, 163
variable names, 83–84
writing distances on the number
    line, 154–55
positive numbers, 10, 11, 155–56
  odd, 10–11
    adding up, 9–10, 13–14,
      17–26
powers, *see* exponentiation;
    exponents
prime factors, 360–62, 363
prime numbers, 360–62
proportions, 399–400
pyramid problem, 42–46, 56
Pythagorean theorem, 49–56,
    99–100, 161
  deck problem and, 62–63, 65–66
  equation solving rules and,
    107–15
  exam questions on, 73–74
  in geometric meaning of
    expressions, 101–3
  proving, 104–6
  in scale metaphor, 106
Pythagorean triples, 51–53
Pythagoras, 51

quadratic equations, 342–44
  exam questions on, 411–12
  factoring and, 387
  solving, 383–89
  standard form of, 383–84
  visualizing solution of, 343–44

with all terms, 386
with no constant term, 385, 386
with no linear term, 384–85, 386
quadratic formula, 387–88
quick and dirty tips, 6

range, 166–67
rational expressions, 273–74
rational numbers, 157–60, 162, 163
real numbers, 162–63
right angle, 43–44
roots, 248, 252–54
    exam questions on, 262
    exponentiation and, 250–51, 254–56
    summary of properties of, 253
roots of polynomials, 344–59
    exam questions on, 410–11
    imaginary, 351–53
    multiplicity of, 354–55

secret agent math-libs, 5
    decrypting message, 311–14
    hiding keys, 117–22
    is math boring?, 14–16
    mystery message, 257–59
    nightmare problem, 389–99
    Pythagoras' message, 53–56
    safe commute, 150–53
shopping problem, 34–37, 38–40
slope, 291–92
    calculating, 292–94
    in equations of lines, 300–310
    of parallel and perpendicular lines, 318–20
square, how to make, 58–62
square roots, 46–49, 357
    calculating, 47–49
    exam questions on, 73
    in golden rule of equation solving, 109
    in isolating variable in equations, 113
    of 2, 161–62
squaring numbers, 19, 46, 241
    exam questions on, 29
    perfect squares, 18–26, 46
    exponents and, 19–20
subtraction, 11, 199
    of fractions, 202–4
    in golden rule of equation solving, 108–9
    of like terms, 236–37
    in moving variable to one side of equation, 112–13, 169–70
    of negative numbers, 212–14

in order of operations, 90–91
of polynomials, 298–99
with roots, 252–53
symbol for, 12
of variables, 214–15
symbols (notation), 12, 77–78, 81
    exam questions on, 28–29

terms, 215, 266
    like, 234–35
        combining, 234–40, 267–68
triangles:
    hypotenuse of, 49–50, 65–66, 74, 161
    Pythagorean theorem and, see Pythagorean theorem
trinomials, 269–70
tutorials, see algebra tutorials

units, converting, 399–400

variables, 12, 38–39, 77, 81, 82–86, 100
    absolute value of, 145–47
    adding and subtracting, 214–15
    calculating distance between, 153–54
    division of, 219
    exam questions on, 130
    in expressions, 86–87, 97
    isolating in equation, 113, 170–71
    moving to one side of equation, 112–13, 169–70
    multiplication of, 216–19
    naming, 82–84
    two, linear equations with, 191–93
    what can be done with, 84–85
    working with, 85–86
    x, 81, 82, 83
volume, 218

Watch Out, 6
    absolute values, 150
    distributive property, 228
    dividing by zero, 233–34
    integers and rational numbers, 160
    linear equations, 164–65
    multiplying and dividing inequalities by negative numbers, 184
    multiplying with exponents, 244
    negative numbers and PEMDAS, 93
    quadratic equations, 386
    quadratic formula, 388

Watch Out (*cont'd*)
   roots, 253
   "small numbers," 144
   square roots, 48
   zero, 156

x, 81, 82, 83
x-axis, 65

x-y plane (coordinate plane), 65
   plotting points on, 67–68

y-axis, 65

zero, 155, 156
   dividing by, 233–34
   as exponent, 246–47

ABOUT THE AUTHOR

When not writing and hosting *The Math Dude's Quick and Dirty Tips to Make Math Easier* podcast, Jason Marshall works as a research scientist at the California Institute of Technology (Caltech) studying the infrared light emitted by starburst galaxies and quasars. Prior to that, he was a postdoctoral scholar at NASA's Jet Propulsion Laboratory (JPL). Jason obtained a Ph.D. from Cornell University, where he also taught many physics and astronomy classes, and a B.S. in Astrophysics from UCLA. In addition to these astronomical interests, Jason also enjoys many earthly diversions. These include travel, technology, green building, soccer, and spending time with his wife, Shannon, and their brown tabby cat in and around their Los Angeles area home.

Quick and Dirty Tips™

Helping you do things better.

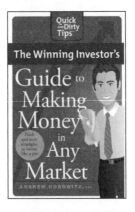